Certified SOLIDWORKS Professional
Advanced Preparation Materials
SOLIDWORKS 2024

T0186224

SDC Publications
P.O. Box 1334
Mission, KS 66222
913-262-2664
www.SDCpublications.com
Publisher: Stephen Schroff

ISBN-13: 978-1-63057-643-1
ISBN-10: 1-63057-643-3

Printed and bound in the United States of America.

Acknowledgments

Thanks as always to my wife Vivian and my daughter Lani for always being there and providing support and honest feedback on all the chapters in the textbook.

Additionally, thanks to Kevin Douglas and Peter Douglas for writing the forewords.

I also have to thank SDC Publications and the staff for their continuing encouragement and support for this edition of **SOLIDWORKS 2024 CSWPA Preparation Materials**. Thanks also to Tyler Bryant for putting together such a beautiful cover design.

Finally, I would like to thank you, our readers, for your continued support. It is with your consistent feedback that we were able to create the lessons and exercises in this book with more detailed and useful information.

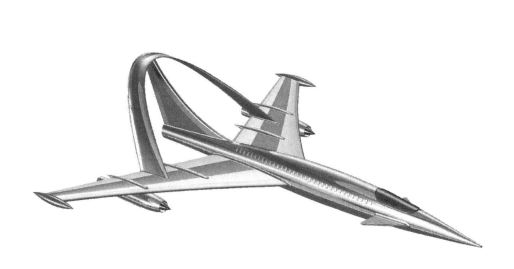

Foreword

For more than two decades, I have been fortunate to have worked in the fast-paced, highly dynamic world of mechanical product development providing computer-aided design and manufacturing solutions to thousands of designers, engineers, and manufacturing experts in the western US. The organization where I began this career was US CAD in Orange County, CA, one of the most successful SOLIDWORKS Resellers in the world. My first several years were spent in the sales organization prior to moving into middle management and ultimately President of the firm. In the mid-1990s is when I met Paul Tran, a young, enthusiastic instructor who had just joined our team.

Paul began teaching SOLIDWORKS to engineers and designers of medical devices, automotive and aerospace products, high tech electronics, consumer goods, complex machinery and more. After a few months of watching him teach and interacting with students during and after class, it was becoming pretty clear – Paul not only loved to teach, but his students were the most excited with their learning experience than I could ever recall from previous years in the business. As the years began to pass and thousands of students had cycled through Paul's courses, what was eye opening was Paul's continued passion to educate as if it were his first class and students in every class, without exception, loved the course.

Great teachers not only love their subject, but they love to share that joy with students – this is what separates Paul from others in the world of SOLIDWORKS Instruction. He has always gone well beyond learning the picks & clicks of using the software, to best practice approaches to creating intelligent, innovative, and efficient designs that are easily grasped by his students. This effective approach to teaching SOLIDWORKS has translated directly into Paul's many published books on the subject. His latest effort with SOLIDWORKS 2024 is no different. Students that apply the practical lessons from basics to advanced concepts will not only learn how to apply SOLIDWORKS to real world design challenges more quickly but will gain a competitive edge over others that have followed more traditional approaches to learning this type of technology.

As the pressure continues to rise on U.S. workers and their organizations to remain competitive in the global economy, raising not only education levels but technical skills is paramount to a successful professional career and business. Investing in a learning process towards the mastery of SOLIDWORKS through the tutelage of the most accomplished and decorated educator and author in Paul Tran will provide a crucial competitive edge in this dynamic market space.

Kevin Douglas
Vice President Sales/Board of Advisors, GoEngineer Inc.

Preface

I first met Paul Tran when I was busy creating another challenge in my life. I needed to take a vision from one man's mind, understand what the vision looked like, how it was going to work and comprehend the scale of his idea. My challenge was I was missing one very important ingredient, a tool that would create a picture with all the moving parts.

Research led me to discover a great tool, SOLIDWORKS. It claimed to allow one to make 3D components, in picture quality, on a computer, add in all moving parts, assemble it and make it run, all before money was spent on bending steel and buying parts that may not fit together. I needed to design and build a product with thousands of parts, make them all fit and work in harmony with tight tolerances. The possible cost implications of failed experimentation were daunting.

To my good fortune, one company's marketing strategy of selling a product without an instruction manual and requiring one to attend an instructional class to get it, led me to meet a communicator who made it all seem so simple.

Paul Tran has worked with and taught SOLIDWORKS as his profession for 35 years. Paul knows the SOLIDWORKS product and manipulates it like a fine musical instrument. I watched Paul explain the unexplainable to baffled students with great skill and clarity. He taught me how to navigate the intricacies of the product so that I could use it as a communication tool with skilled engineers. *He teaches the teachers*.

I hired Paul as a design engineering consultant to create thousands of parts for my company's product. Paul Tran's knowledge and teaching skill have added immeasurable value to my company. When I read through the pages of these manuals, I now have an "instant replay" of his communication skill with the clarity of having him looking over my shoulder - *continuously*. We can now design, prove and build our product and know it will always work and not fail. Most important of all, Paul Tran helped me turn a blind man's vision into reality and a monument to his dream.

Thanks Paul.

These books will make dreams come true and help visionaries change the world.

Peter J. Douglas
CEO, Cake Energy, LLC

Images courtesy of C.A.K.E. Energy Corp., designed by Paul Tran

Author's Note

CSWP - Advanced Preparation Materials is comprised of lessons based on the feedback from Paul's former CSWP students and engineering professionals. Paul has more than 35 years of experience in the fields of mechanical and manufacturing engineering; 2/3 of those years were spent teaching and supporting the SOLIDWORKS software and its add-ins. As an active Sr. SOLIDWORKS instructor and design engineer, Paul has worked and consulted with hundreds of reputable companies including IBM, Intel, NASA, US-Navy, Boeing, Disneyland, Medtronic, Edwards Lifesciences, Oakley, Kingston, Community Colleges, Universities, and many others. Today, he has trained more than 14,000 engineering professionals, and given guidance to one-half of the number of Certified SOLIDWORKS Professionals and Certified SOLIDWORKS Experts (CSWP & CSWE) in the state of California.

Every lesson in this book was created based on the actual CSWPA Examinations. Each of these projects has been broken down and developed into easy and comprehendible steps for the reader. Furthermore, every challenge is explained very clearly in short chapters, ranging from 30 to 50 pages. Each and every single step comes with the exact screenshot to help you understand the main concept of each design more easily. Learn the CSWP Advanced Preparation Materials at your own pace, as you progress from Parts, Assemblies, Drawings and then to more complex design challenges.

About the Training Files

The files for this textbook are available for download on the publisher's website at www.SDCpublications.com/downloads/978-1-63057-643-1. They are organized by the chapter numbers and the file names that are normally mentioned at the beginning of each chapter or exercise. In the **Completed Parts** folder you will also find copies of the parts, assemblies and drawings that were created for cross referencing or reviewing purposes.

It would be best to make a copy of the content to your local hard drive and work from these documents; you can always go back to the original training files location at any time in the future, if needed.

Who this book is for

This book is for the mid-level to advanced user, who is already familiar with the SOLIDWORKS program. To get the most out of this CSWPA-Certification Preparation book it is strongly recommended that you have studied and completed all the lessons in the Basic and Advance textbooks. It is also a great resource for the more CAD literate individuals who want to expand their knowledge of the different features that SOLIDWORKS 2024 has to offer.

The organization of the book

The chapters in this book are organized in the logical order in which you would learn the advanced SOLIDWORKS 2024 topics. Each chapter will guide you through some different tasks, from designing or repairing a mold, to developing a complex sheet metal part; from some 3D sketch for weldments to advancing through more complex tasks that are common to all surface modeling challenges. You will also learn to work with part and assembly drawings and managing different levels of the Bill of Materials.

The conventions in this book

This book uses the following conventions to describe the actions you perform when using the keyboard and mouse to work in SOLIDWORKS 2024:

Click: means to press and release the mouse button. A click of a mouse button is used to select a command or an item on the screen.

Double Click: means to quickly press and release the left mouse button twice. A double mouse click is used to open a program or show the dimensions of a feature.

Right Click: means to press and release the right mouse button. A right mouse click is used to display a list of commands, a list of shortcuts that is related to the selected item.

Click and Drag: means to position the mouse cursor over an item on the screen and then press and hold down the left mouse button; still holding down the left button, move the mouse to the new destination and release the mouse button. Drag and drop makes it easy to move things around within a SOLIDWORKS document.

Bolded words: indicates the action items that you need to perform.

Italic words: Side notes and tips that give you additional information, or to explain special conditions that may occur during the course of the task.

Numbered Steps: indicates that you should follow these steps in order to successfully perform the task.

Icons: indicates the buttons or commands that you need to press.

SOLIDWORKS 2024

SOLIDWORKS 2024 is a program suite, or a collection of engineering programs that can help you design better products faster. SOLIDWORKS 2024 contains different combinations of programs; some of the programs used in this book may not be available in your suites.

Start and exit SOLIDWORKS

SOLIDWORKS allows you to start its program in several ways. You can either double click on its shortcut icon on the desktop or go to the Start menu and select the following: All Programs / SOLIDWORKS 2024 / SOLIDWORKS or drag a SOLIDWORKS document and drop it on the SOLIDWORKS shortcut icon.

Before exiting SOLIDWORKS, be sure to save any open documents, and then click File / Exit; you can also click the X button on the top right of your screen to exit the program.

Using the Toolbars

You can use toolbars to select commands in SOLIDWORKS rather than using the drop down menus. Using the toolbars is normally faster. The toolbars come with commonly used commands in SOLIDWORKS, but they can be customized to help you work more efficiently.

To access the toolbars, either right click in an empty spot on the top right of your screen or select View / Toolbars.

To customize the toolbars, select Tools / Customize. When the dialog pops up, click on the Commands tab, select a category, then drag an icon out of the dialog box and drop it on a toolbar that you want to customize. To remove an icon from a toolbar, drag an icon out of the toolbar and drop it into the dialog box.

Using the task pane

The task pane is normally kept on the right side of your screen. It displays various options like SOLIDWORKS resources, Design library, File explorer, Search, View palette, Appearances and Scenes, Custom properties, Built-in libraries, Technical alerts, and news, etc.

The task pane provides quick access to any of the mentioned items by offering the drag and drop function to all of its contents. You can see a large preview of a SOLIDWORKS document before opening it. New documents can be saved in the task pane at any time, and existing documents can also be edited and re-saved. The task pane can be resized, closed, or moved to different locations on your screen if needed.

TABLE OF CONTENTS

CSWPA – Drawing Tools

CSWPA – Mold Making

CSWPA – Weldments

Chapter 3: Advanced Weldments — 3-1

CSWPA – Sheet Metal

Chapter 4: Advanced Sheet Metal   4-1

CSWPA – Surfacing

Chapter 5: Advanced Surfacing 5-1

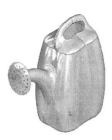

Glossary and Index

SOLIDWORKS 2024 Quick-Guides:

Quick Reference Guide to SOLIDWORKS 2024 Command Icons and Toolbars.

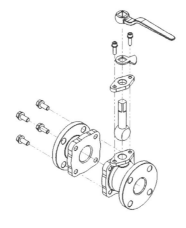

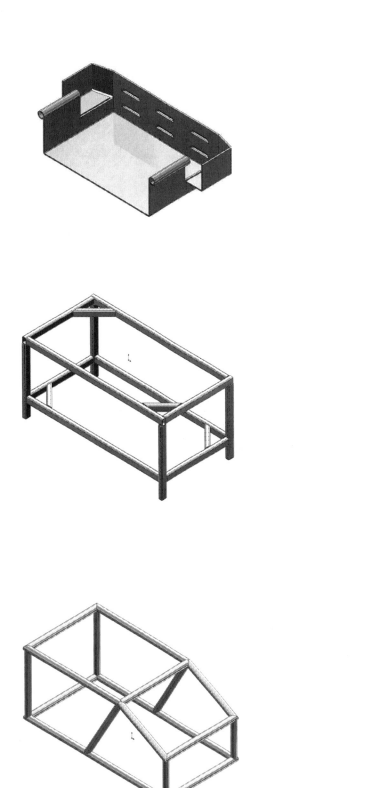

CHAPTER 1

CSWP – Advanced Drawing Tools

CSWP – Advanced Drawing Tools

Certified SOLIDWORKS Professional Advanced Drawing Tools

The completion of the Certified SOLIDWORKS Professional Advanced Drawing Tools (CSWPA-DT) exam proves that you have successfully demonstrated your ability to use the tools found in the SOLIDWORKS Drawing environment.

Employers can be confident that you understand the tools and functionality to create engineering drawings using SOLIDWORKS.

Note: You must use at least SOLIDWORKS 2010 for this exam. Any use of a previous version will result in the inability to open some of the testing files.

Exam Length: 100 minutes

Minimum Passing grade: 75%

Re-test Policy: There is a minimum 14-day waiting period between every attempt of the CSWPA-DT exam. Also, a CSWPA-DT exam credit must be purchased for each exam attempt.

All candidates receive electronic certificates and a personal listing on the CSWP directory when they pass.

Exam features hands-on challenges in many of these areas of SOLIDWORKS drawing functionality such as:

Basic View Creation, Section Views, Auxiliary Views, Alternate position Views, Broken Out Sections, Lock View/Sheet Focus, Transferring Sketch Entities to/from Views, Bill of materials, and Custom Properties.

CSWP – Advanced Drawing Tools

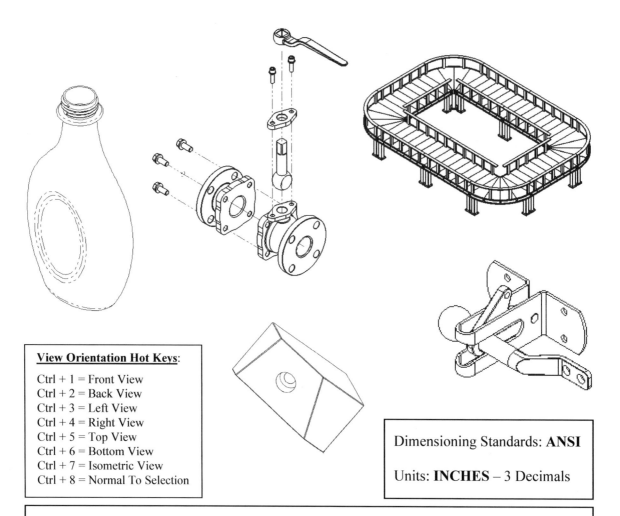

View Orientation Hot Keys:

Ctrl + 1 = Front View
Ctrl + 2 = Back View
Ctrl + 3 = Left View
Ctrl + 4 = Right View
Ctrl + 5 = Top View
Ctrl + 6 = Bottom View
Ctrl + 7 = Isometric View
Ctrl + 8 = Normal To Selection

Dimensioning Standards: **ANSI**

Units: **INCHES** – 3 Decimals

Tools Needed:

Part Template	Assembly Template	Drawing Template
View Palette	Section View	Named View
Measure	Auto Balloon	Bill of Materials

CHALLENGE 1

1. Opening a part document: (This challenge focuses on the drawing view creation and calculating the surface perimeters).

Select **File / Open**.

Browse to the Training Folder and open a part document named: **Plastic Bottle.sldprt**.

2. Transferring to a drawing:

Select **File / Make Drawing From Part** (arrow).

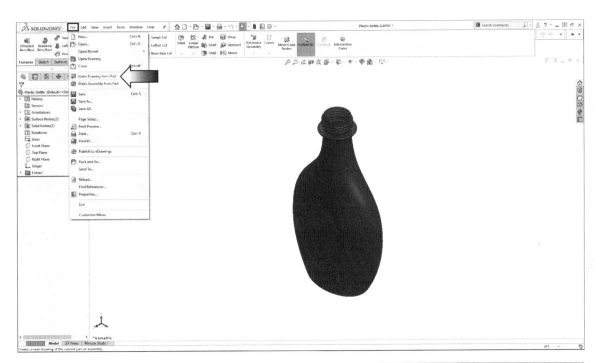

Select the default **Drawing** template.

Click **OK**.

The drawing paper size will be changed in the next step.

3. Changing the paper size:

Right-click inside the drawing and select **Properties**.

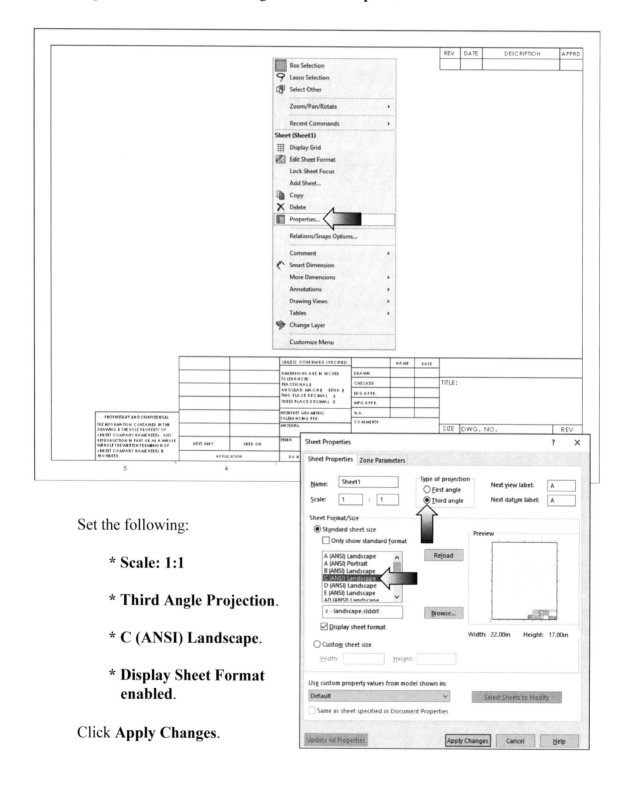

Set the following:

    * Scale: 1:1

    * Third Angle Projection.

    * C (ANSI) Landscape.

    *** Display Sheet Format
    enabled**.

Click **Apply Changes**.

4. Adding the drawing views:

Expand the **View Palette** (arrow) and drag the **Front** View into the drawing approximately as shown.

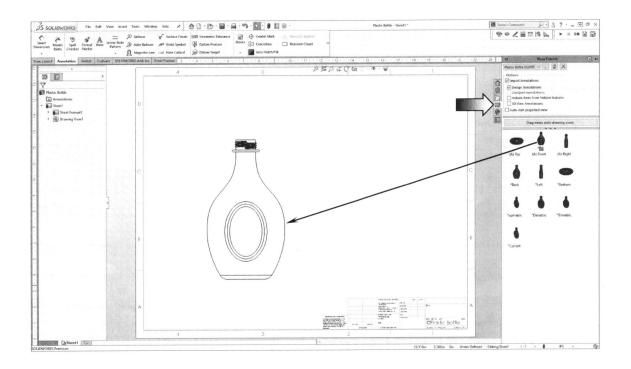

Create an Isometric View by projecting from the Front view, or by dragging and dropping from the View Palette.

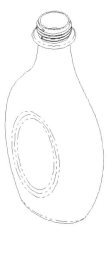

Place the Isometric view on the right side of the Front view.

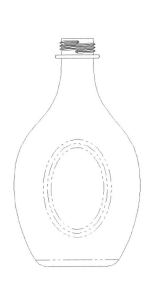

For clarity, change the tangent edges to With-Font (right click the view's border and select Tangent Edges With Font).

5. Creating a section view:

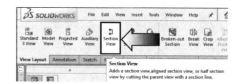

Switch to the **View Layout** tab.

Click the **Section View** command.

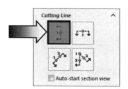

For Cutting Line, select the **Vertical** option (arrow).

Place the Cutting Line in the <u>center</u> of the Front view and click the **green** check mark (arrow) to accept the line placement.

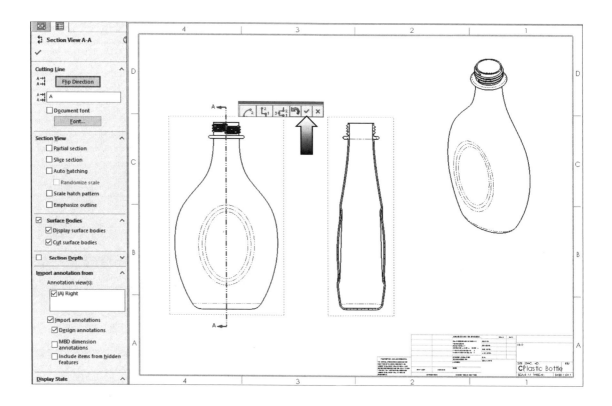

Click Flip Direction if needed, the Section Arrows should be pointing to the <u>left</u>.

Place the Section view on the right side of the Front view.

Move the Isometric view to the upper right hand side. This view is for reference use only.

6. Measuring the surface Perimeter:

Zoom in on the section view; we will need to select the sectioned surface and measure its Perimeter.

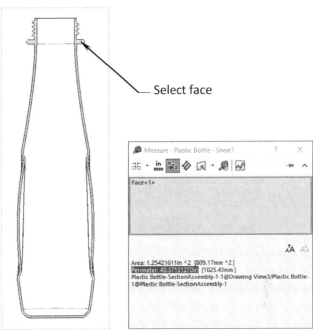

Switch to the **Evaluate** tab and click **Measure**.

Locate the **Perimeter** value and enter it here:

_____ inches.

7. Creating an aligned section view:

Double-click the dotted border of the Front view to lock it.

The **Lock View Focus** option allows you to add sketch entities to a view so that that when the view is moved, the entities will move with the view. This works well for adding section lines manually.

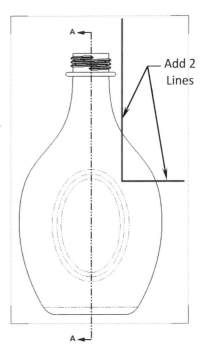

Switch to the **Sketch** tab and sketch **2 Lines** as shown in the image on the right.

Add the vertical and horizontal dimensions to fully define the sketch. (Multiple lines are often used to create an Aligned Section View.)

Hold the **Control** key and select the <u>Vertical Line 1st</u>, and then select the <u>Horizontal Line after</u>.

Switch to the **View Layout** tab and select the **Section View** command (arrow).

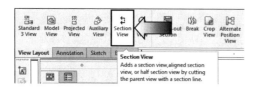

An **Aligned Section View** is created and labeled as **Section B-B**.

Ensure that the Direction Arrows match the image shown below.

Click the **Flip Direction** button if needed (arrow).

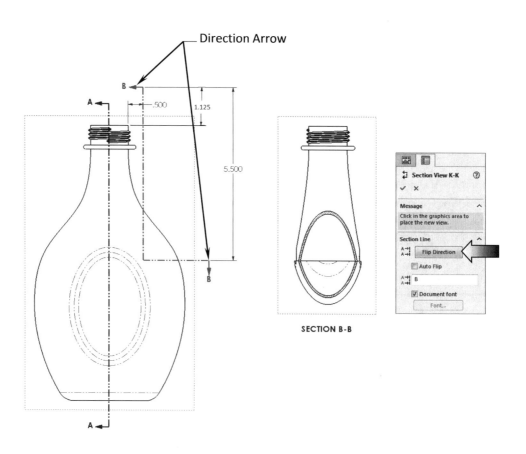

Direction Arrow

SECTION B-B

8. Measuring the surface Perimeter:

Zoom in on the section view; we will need to select the upper surface of the Section B-B and measure its perimeter.

Switch to the **Evaluate** tab and click the **Measure** command.

Select only the upper <u>face</u> as noted.

Locate the **Perimeter** measurement and enter it here:

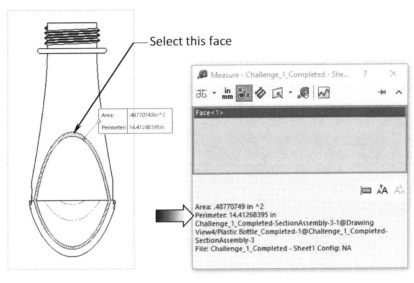

_____ inches.

9. Saving your work:

Select **File / Save As**.

Enter **Challenge_1.slddrw** for the file name.

Click **Save**.

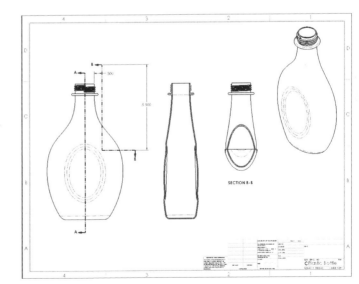

Summary:

The key features to the Challenge 1 are:

Creating the Section Views and Measuring the Perimeter of the sectioned surfaces.

CHALLENGE 2

1. Opening an assembly document: (This challenge focuses on the orientation modifications and drawing view creation).

Select **File / Open**.

Browse to the Training Folder and open an assembly document named: **Latch Assembly.sldasm**.

In this Challenge, the orientation of the assembly has been changed to some oblique angle. You will need to come up with a way to find the correct angle and change the orientation back to normal prior to making the drawing.

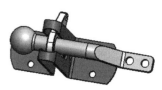

Top View   Isometric View

Change to different view Orientations such as the Front, Top, Right, and Isometric view to examine the default orientations of this assembly.

The Top view will be used to correct the orientation of the assembly.

Front View   Right View

Select the component **Base SM** and click **Edit Component**.

Select face

Rotate the assembly and open a **new sketch** on the <u>upper face</u> as noted.

2. Creating the 1ˢᵗ reference sketch:

We will need to rotate the Base_SM to the horizontal position. There are several methods to find the current angle of the Base but we will go with creating a reference sketch approach.

Sketch a **Construction Line** as indicated.

(Press **Control+8** to rotate normal to the sketch face.)

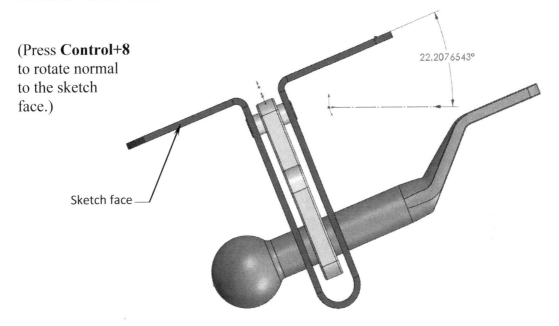

Add a <u>reference</u> **angular dimension** as shown above.

Change the number of decimals to **5 places**. This angular dimension will be used to rotate the entire assembly to the correct orientation as shown in the sample image below.

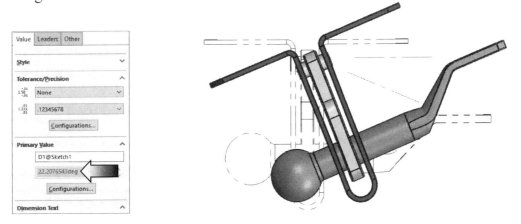

3. Modifying the Arrow Keys angle:

Select **Tools, Option, System Options**.

Select the **View** option and change the angle of the **Arrow Keys** to **22.20765**.

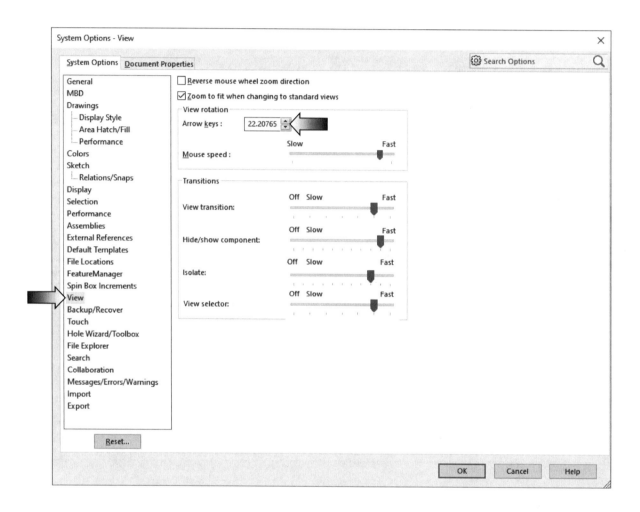

Click **OK**.

At this point, if any of the arrow keys are pressed, the model will rotate precisely 22.20765° each time. To rotate "normal to screen" hold down the Alt key while pressing one of the 4 arrow keys.

Exit the sketch.

4. Changing the Top view orientation:

Remain in the **Top** view orientation.
Hold down the **Alt** key and press the **Left Arrow** key <u>once</u>.

The entire assembly is rotated 22.20765° clockwise.
This is the correct orientation for
the top view.

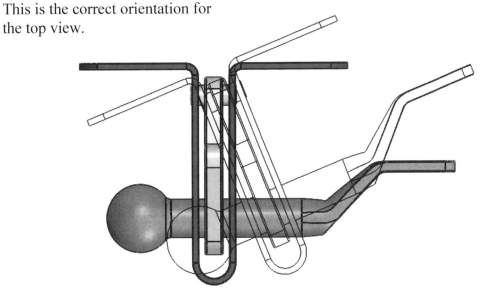

5. Creating the 2nd reference sketch:

Select the <u>face</u> on the right side of the
Base_SM and open a **new sketch**.

Press **Control+8** to rotate normal
to the sketch face. This is the
current Right view orientation.

Sketch a horizontal **centerline**
and add a <u>reference</u> **angular
dimension**.

<u>Exit</u> the sketch.

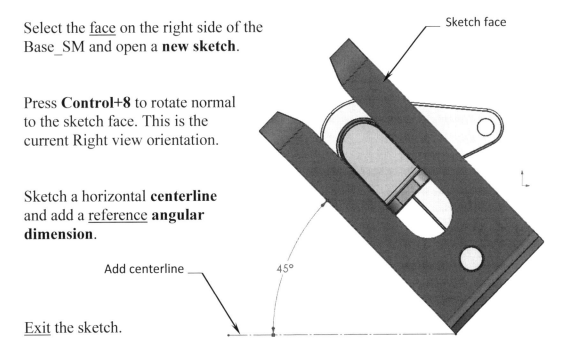

Sketch face

Add centerline

45°

6. Modifying the Arrow Keys angle:

Select **Tools, Option, System Options**.

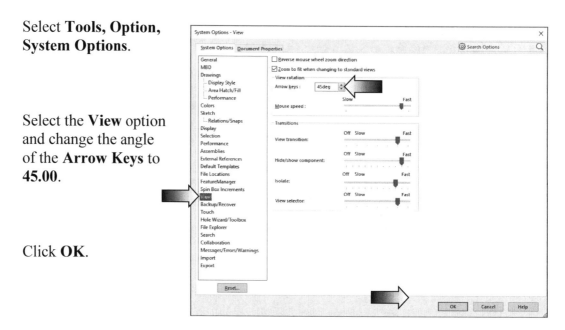

Select the **View** option and change the angle of the **Arrow Keys** to **45.00**.

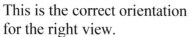

Click **OK**.

7. Changing the Right view orientation:

Hold down the **Alt** key and press the **Right Arrow** key <u>once</u>.

The entire assembly is rotated 45° counterclockwise.
This is the correct orientation
for the right view.

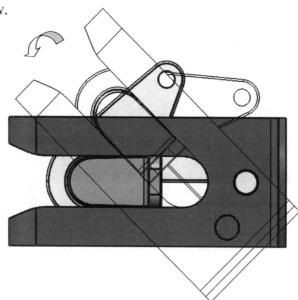

8. Saving a new named view:

Custom views can be created and saved in the model or in an assembly document so that they can be displayed in a drawing at a later time.

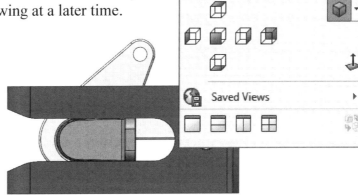

The views are saved in the Orientation dialog and get carried over to the drawing and listed on the Properties tree.

Press the **Spacebar** to access the Orientation dialog.

Click the **New View** button .

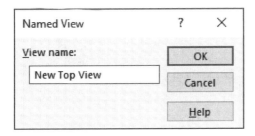

Enter: **New Top View** in the Named View dialog and press **OK**.

The new view is saved and displayed in the Orientation dialog.

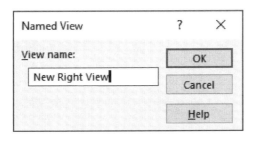

Repeat the last step and save another Named View called: **New Right View**.

Click **OK** and save the assembly document.

9. Making a drawing from assembly:

Select **File / Make Drawing from Assembly** (arrow).

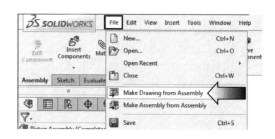

Select the **Drawing** template.

Click **OK**.

Right-click inside the drawing and select **Properties**.

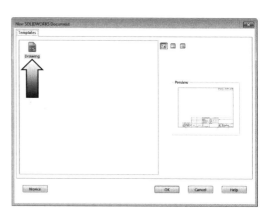

Change the paper size to **C-Landscape**.

Change the **Scale** to **1:1.**

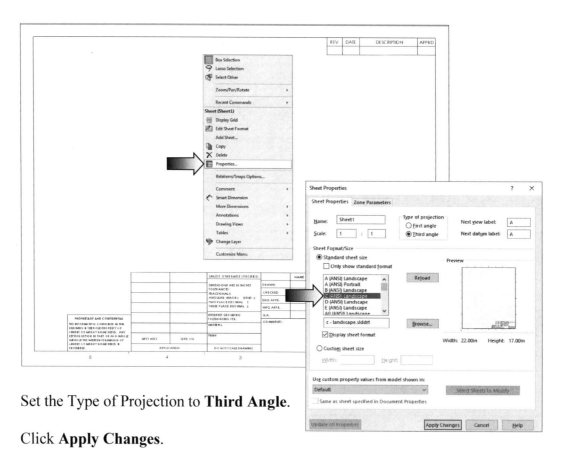

Set the Type of Projection to **Third Angle**.

Click **Apply Changes**.

10. Adding the first drawing view:

Drag and drop the **Top** drawing view from the **View Palette**.

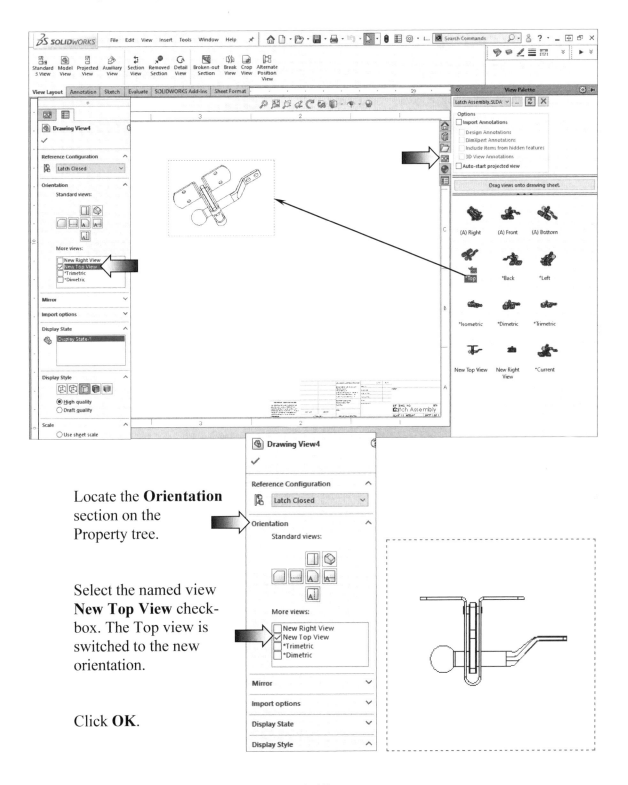

Locate the **Orientation** section on the Property tree.

Select the named view **New Top View** checkbox. The Top view is switched to the new orientation.

Click **OK**.

11. Creating the projected drawing views:

New drawing views can now be projected vertically or horizontally from the new view.

Switch to the **View Layout** tab and click the **Projected View** command.

Click the dotted border of the **Top** view to start the projection.

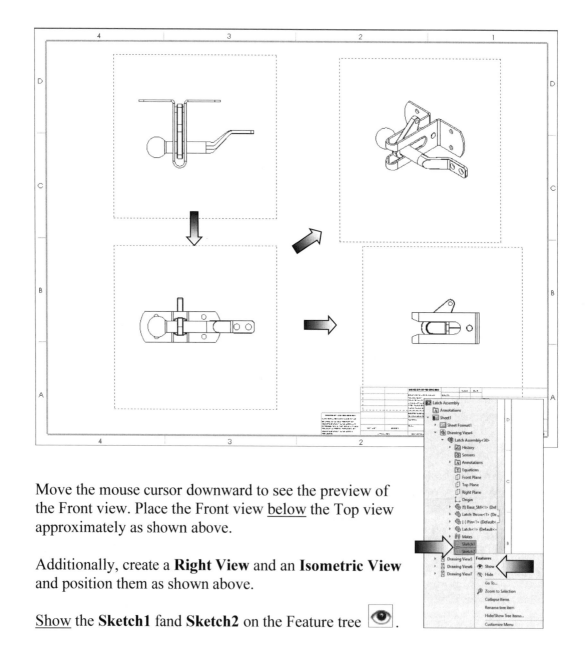

Move the mouse cursor downward to see the preview of the Front view. Place the Front view <u>below</u> the Top view approximately as shown above.

Additionally, create a **Right View** and an **Isometric View** and position them as shown above.

<u>Show</u> the **Sketch1** fand **Sketch2** on the Feature tree 👁.

12. Adding reference lines:

Zoom in on the lower right corner of the drawing.

Right-click anywhere inside the drawing and select **Lock Sheet Focus**.
This forces the new lines to be part of the sheet. If any of the drawing view is moved, the added line will <u>not</u> move.

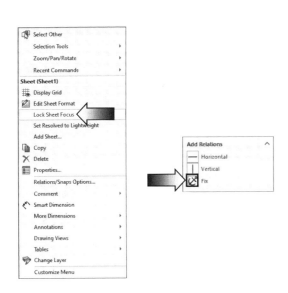

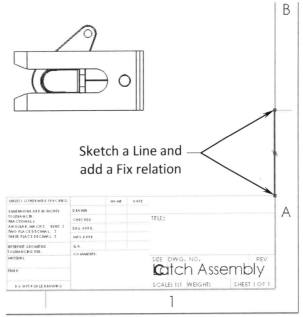

Sketch a Line and add a Fix relation

Sketch a vertical **Line** exactly on the inner border line as shown.

Add a **<u>Fix</u>** relation to the line so that it will not move.

Double-click the dotted border of the <u>right view</u> to **Lock View Focus**. This will force the new line and its dimensions to be part of the view. If the drawing view is moved, the line and its dimension <u>will</u> also move with it.

Sketch a **second Line** and add two dimensions as shown.

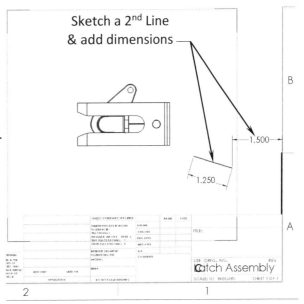

Sketch a 2nd Line & add dimensions

13. Adding an angular dimension:

Create an angular dimension between the 2nd line and the outer edge of the Latch.

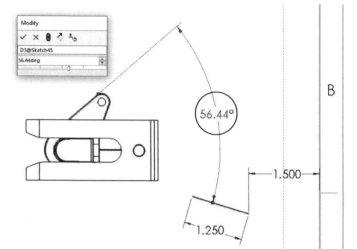

Change the angle to **56.44°**.

14. Finding the value of X:

Add a reference dimension between the left endpoint of the 2nd line and the edge on the right side of the Right drawing view.

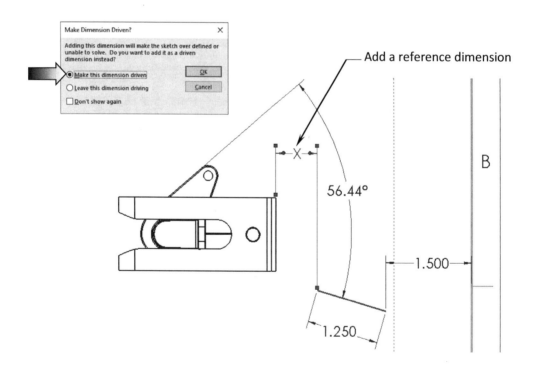

Enter the value of the **dimension X** here: _____ in.

15. Creating a section view:

Double-click the dotted border of the **Right** drawing view to lock the view focus.

Sketch a **Line** and add a **Parallel** relation to the <u>outer edge</u> of the Latch.

Switch to the **View Layout** tab and click **Section View**.

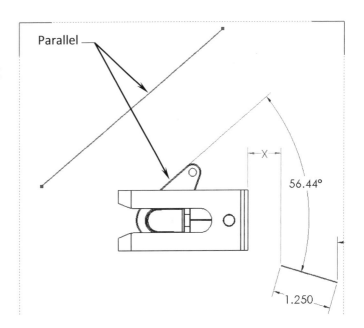

Click **OK** to accept the default options in the **Section Scope** dialog box.

Place the section view approximately as shown.

The direction arrows should be pointing towards the right drawing view.

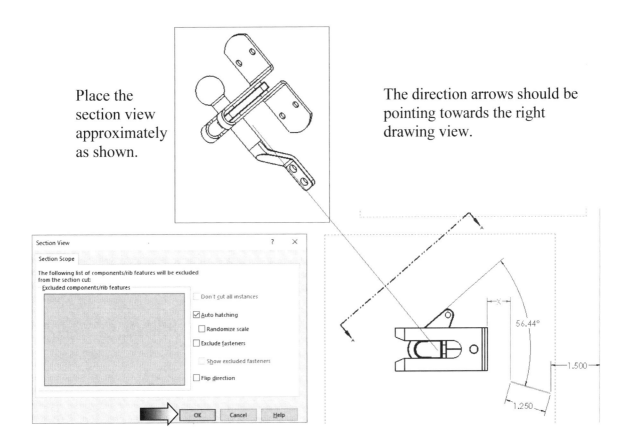

16. Finding the value of Y:

Double-click the dotted border of the **Right** drawing view to lock its view focus.

Convert the horizontal edge of the **Base_SM** to a line (image below).

Add an angular dimension between the edge of the Base_SM and the converted line as shown below.

Click **OK** to accept the reference dimension.

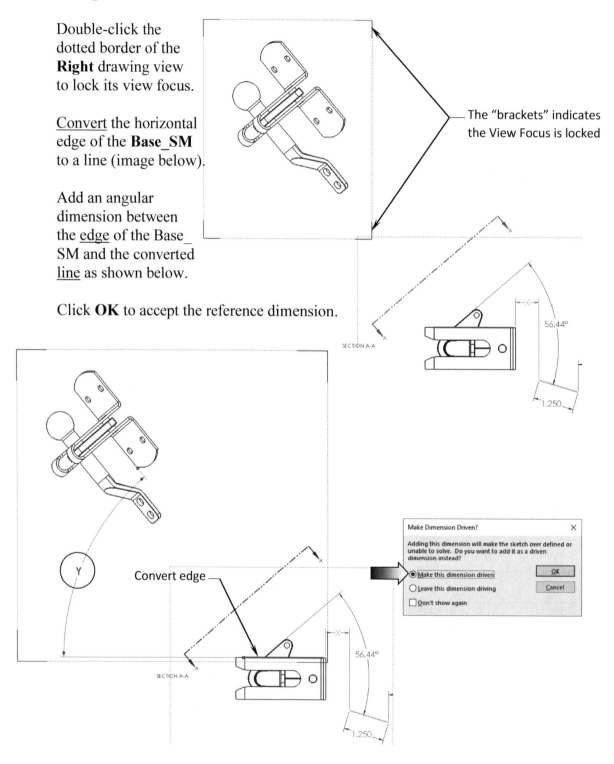

The "brackets" indicates the View Focus is locked

Convert edge

Enter the value of the **dimension Y** here: _____ deg.

17. Creating an Alternate Position view:

The dimensions and the lines in the Right drawing view are hidden for clarity.

Switch to the **View Layout** tab and click: **Alternate Position View**.

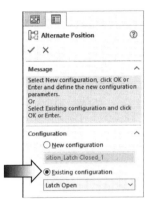

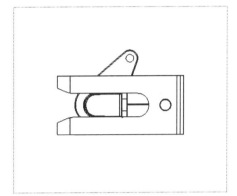

Select the **Right** drawing view.

Click **Existing-Configuration** and select **Latch Open** from the drop-down list.

Click **OK**.

An alternate position view is displayed in phantom line style, on top of the Right drawing view.

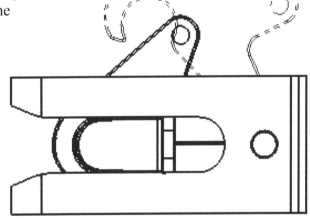

18. Saving your work:

Select **File / Save As**.

Enter **Challenge_2.slddrw** for the file name.

Click **Save**.

Summary:

The key features to the Challenge 2 are:

Creating the drawing views and finding the right orientations and return the views back to normal before creating the drawing views.

Lock and Unlock the View Focus so that reference geometry can be added for measuring and locating other references.

CHALLENGE 3

This challenge examines your skills on the following:

* Creating an assembly drawing.
* Adding balloons.
* Customizing the bill of materials.

1. Opening a drawing document:

Select **File / Open**.

Browse to the Training Folder and open a drawing document named:
Deck Assembly.slddrw.

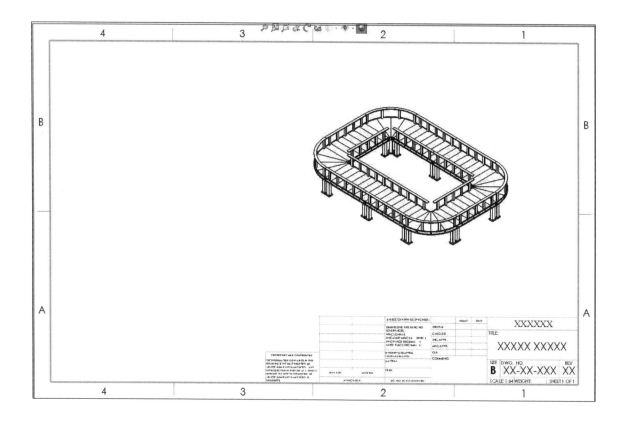

There is only 1 Isometric drawing view in this drawing.
A Bill of Materials and Balloons need to be added to the drawing view.

2. Inserting the balloons:

Balloons are used to identify the item numbers in the bills of materials.

Switch to the **Annotation** tab and click the **Auto Balloon** command . Select the Isometric view's border.

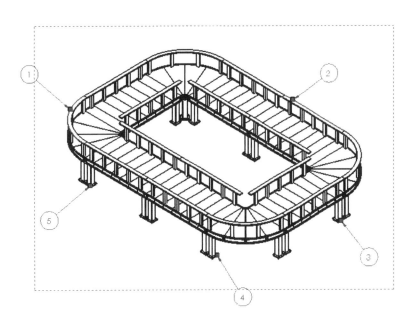

By default, each unique component gets a balloon assigned to it automatically.

Change the balloon settings to **Circular**, **2 characters** and click **OK** (arrows).

The item numbers reflect the order of the components listed in the top level assembly. Changes done to the order of the components in the assembly feature tree will populate the balloons and the bill of materials.

3. Adding a bill of materials:

In an assembly drawing, a bill of materials is created to display the item numbers, quantities, part numbers, and custom properties of the assembly.

The bill of materials should be linked to a drawing view, so click the dotted border of the Isometric view to activate it.

From the **Annotation** tab, select **Tables / Bill of Materials**.

For Table Template, select **BOM-Standard**.

For BOM Type, select **Parts Only**.

For Part Configuration Grouping, select **Display as One Item Number**.

Click **OK**.

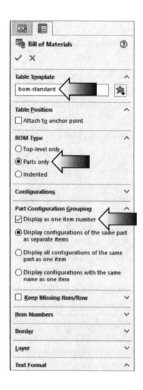

Place the Bill of Materials on the left side of the drawing view.

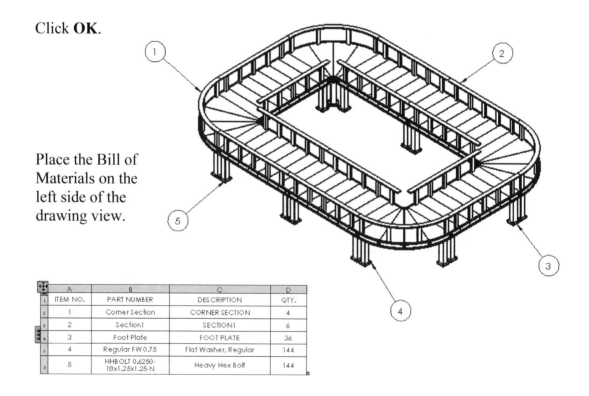

ITEM NO.	PART NUMBER	DESCRIPTION	QTY.
1	Corner Section	CORNER SECTION	4
2	Section1	SECTION1	6
3	Foot Plate	FOOT PLATE	36
4	Regular FW 0.75	Flat Washer, Regular	144
5	HHB OLT 0.6250-18x1.25x1.25-N	Heavy Hex Bolt	144

	A	B	C	D
1	ITEM NO.	PART NUMBER	DESCRIPTION	QTY.
2	1	Corner Section	CORNER SECTION	4
3	2	Section1	SECTION1	6
4	3	Foot Plate	FOOT PLATE	36
5	4	Regular FW 0.75	Flat Washer, Regular	144
6	5	HHBOLT 0.6250-18x1.25x1.25-N	Heavy Hex Bolt	144

Zoom in on the Bill of materials. We will change the Part Number column to include the actual part numbers that were assigned earlier, at the part level.

4. Changing custom properties:

Double-click column header **B** to access the Custom Property options.

Double-click

Column type:
PART NUMBER
CUSTOM PROPERTY
UNIT OF MEASURE
EQUATION
ITEM NO.
PART NUMBER
COMPONENT REFERENCE
TOOLBOX PROPERTY

	A	B	C	D
1	ITEM NO.	PART NUMBER	DESCRIPTION	QTY.
2	1	Corner Section	CORNER SECTION	4
3	2	Section1	SECTION1	6
4	3	Foot Plate	FOOT PLATE	36
5	4	Regular FW 0.75	Flat Washer, Regular	144
6	5	HHBOLT 0.6250-18x1.25x1.25-N	Heavy Hex Bolt	144

Change the Column Type to **Custom Property**.

For Property Name, select **PartNo** from the list.

Column type:
CUSTOM PROPERTY
Property name:
PartNo
Description
PartNo
SW-Author(Author)
SW-Comments(Comments)
SW-Configuration Name(Con
SW-Created Date(Created Dat
SW-File Name(File Name)
SW-Folder Name(Folder Name
SW-Keywords(Keywords)
SW-Last Saved By(Last Saved E
SW-Last Saved Date(Last Save
SW-Long Date(Long Date)
SW-Short Date(Short Date)
SW-Subject(Subject)
SW-Title(Title)

	A		C	D
1	ITEM NO.		DESCRIPTION	QTY.
2	1		CORNER SECTION	4
3	2		SECTION1	6
4	3	01-240-613	FOOT PLATE	36
5	4	01-240-615	Flat Washer, Regular	144
6	5	01-240-614	Heavy Hex Bolt	144

The part numbers for each component are displayed in column B. Adjust the column width by dragging the row divider ⇔‖⇨ .

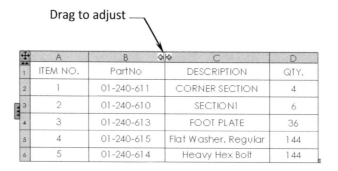

Drag to adjust

5. Adding a Project column:

Right-click the column header **D** and select **Insert / Column Right** (arrow).

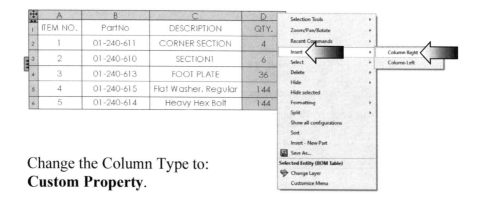

Change the Column Type to: **Custom Property**.

For Property Name, select **Project** from the list.

The project **Deck Design** is displayed in the new column.

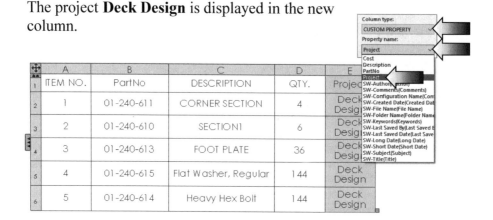

6. Adding a column for Cost:

Right-click the column header **E** and select **Insert / Column Right** (arrow).

Change the Column Type to **Custom Property**.

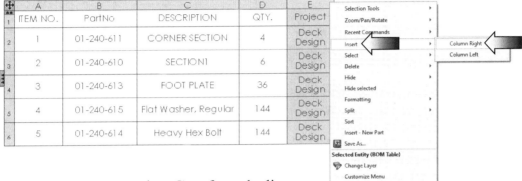

For Property Name, select **Cost** from the list.
The cost for each component is displayed in the new column.

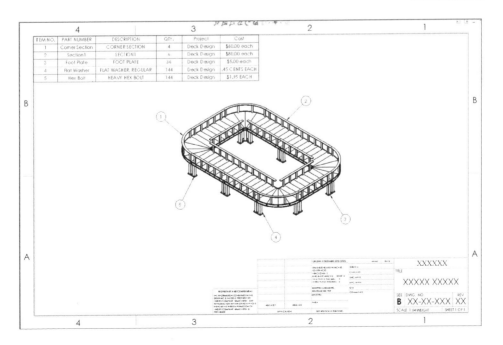

7. Creating a linked note:

Switch to the **Annotation** tab and click **Note**.

Click on the <u>upper face</u> of one of the handrails to attach the note.

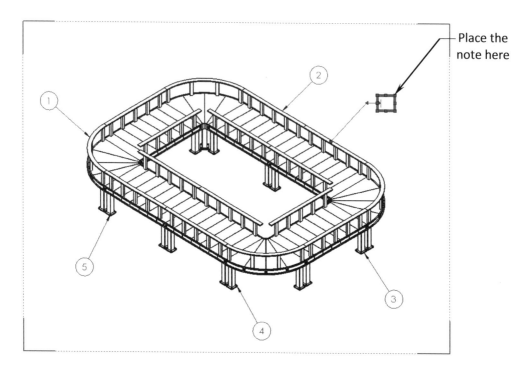

Click the **Link to Property** button.

Under Use Custom Property From, select the **Model Found Here** option.

For Property Name, select **Finish**. The text of the note reads:
Waterproof Coating, Natural.

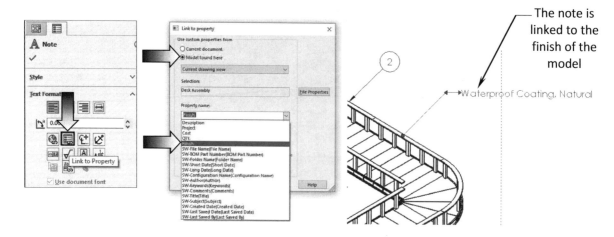

8. Saving your work:

Select **File, Save As**.

Enter **Challenge_3** for the file name.

Click **Save**.

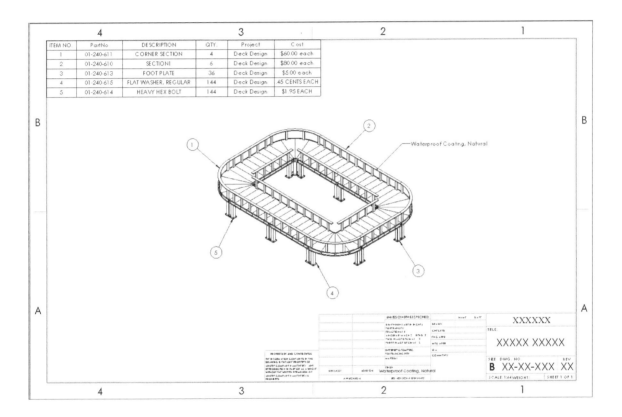

ITEM NO.	PartNo	DESCRIPTION	QTY.	Project	Cost
1	01-240-611	CORNER SECTION	4	Deck Design	$60.00 each
2	01-240-610	SECTION1	6	Deck Design	$80.00 each
3	01-240-613	FOOT PLATE	36	Deck Design	$5.00 each
4	01-240-615	FLAT WASHER, REGULAR	144	Deck Design	45 CENTS EACH
5	01-240-614	HEAVY HEX BOLT	144	Deck Design	$1.95 EACH

Waterproof Coating, Natural

Summary:

The key features to the Challenge 3 are:

Creating an assembly drawing
Adding balloons
Inserting a Bill of materials
Changing the custom properties
Adding new columns
Creating a linked note

CHALLENGE 4

This challenge examines your skills on the following:

* Hide/Show components in a drawing
* Adding projection dimensions

1. Opening a drawing document:

Select **File / Open**.

Browse to the Training Folder and open a drawing document named:
Water Valve Assembly.slddrw.

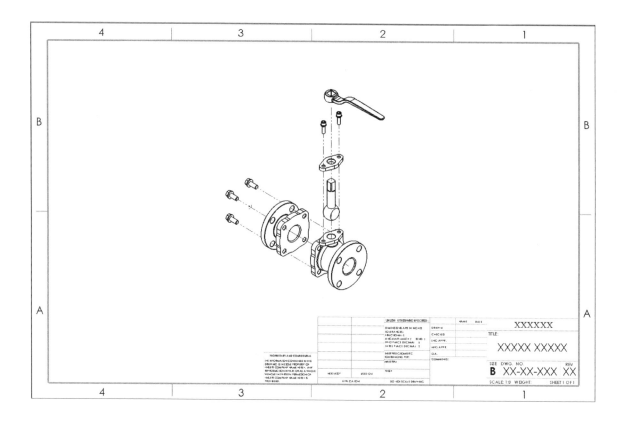

There are some components hidden from the drawing view.
Determine which components are hidden.

2. Finding the hidden components:

Right-click on the dotted border of the Isometric view and select **Properties**.

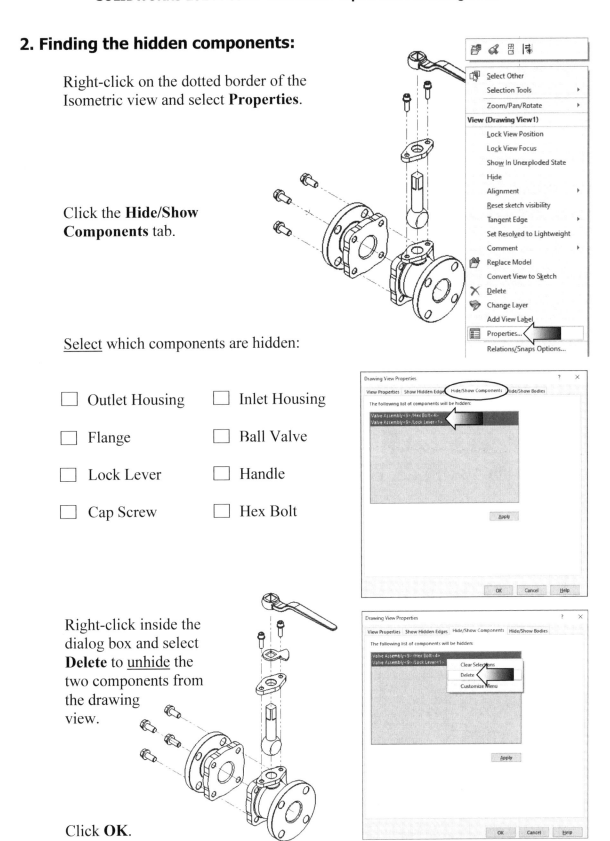

Click the **Hide/Show Components** tab.

<u>Select</u> which components are hidden:

☐ Outlet Housing   ☐ Inlet Housing

☐ Flange   ☐ Ball Valve

☐ Lock Lever   ☐ Handle

☐ Cap Screw   ☐ Hex Bolt

Right-click inside the dialog box and select **Delete** to <u>unhide</u> the two components from the drawing view.

Click **OK**.

3. Adding an angular dimension:

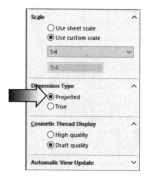

Select the Isometric view's dotted border and change the Dimension Type to **Projected** (arrow).

Add an angular dimension between the <u>edges</u> of the **Handle** and the **Flange**.

Enter the angle value here:_____deg.

4. Saving your work:

Select **File, Save As**.

Enter **Challenge_4** for the file name.

Click **Save**.

Summary:

The key features to the Challenge 4 are:

Hide/Show the hidden components.

Add a Projected dimension for inspection.

CHALLENGE 5

This challenge examines your skills on the following:

* Creating a Broken Out Section view
* Inserting the model dimensions
* Measuring the perimeter of a surface

1. Opening a drawing document:

Select **File / Open**.

Browse to the Training Folder and open a drawing document named:
Helidrone.slddrw.

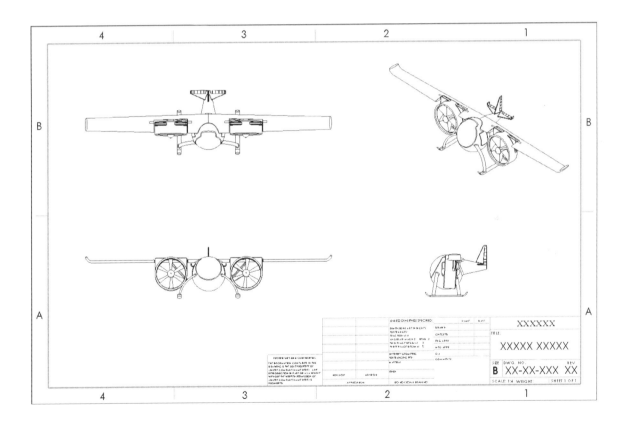

Thera are 4 standard views in this drawing, the Front, Top, Right, and Isometric views. The Top drawing view will be used to create the Broken-Out Section view.

2. Creating a Broken-Out Section view:

Sketch a **Circle** on the Top drawing view and add the 3 dimensions as shown in the image below.

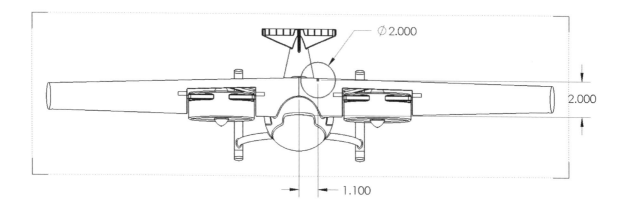

Switch to the **View Layout** tab and click **Broken Out Section**.

Click **OK** in the **Section Scope** dialog box.

For Section Depth, select the top edge of the tail as noted below.

Click **OK**.

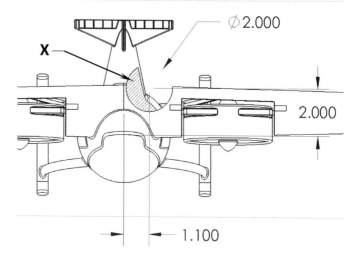

Select edge ———

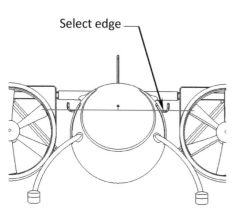

Enter the **Perimeter** of X here: _____.

3. Inserting the Model Dimensions:

Select the <u>Top</u> drawing view and click **Model Items** on the **Annotation** tab.

Import dimensions only to the Top drawing view using these options: **Entire Model, Marked for Drawing.**

Click **OK.**

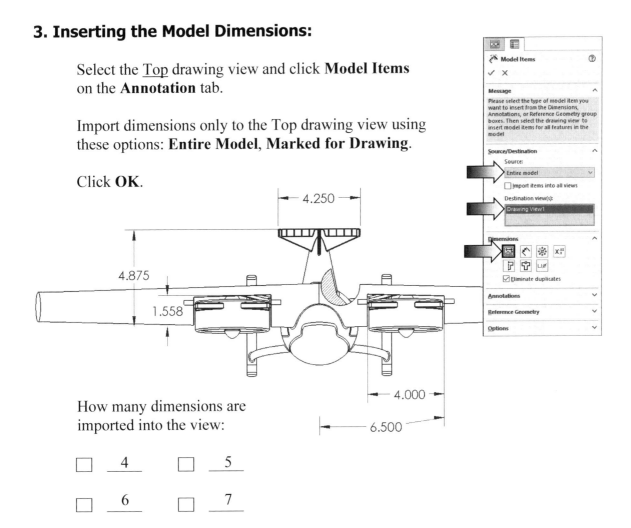

How many dimensions are imported into the view:

☐   4   ☐   5

☐   6   ☐   7

4. Saving your work:

Select **File, Save As.**

Enter **Challenge_5** for the file name.

Click **Save.**

Summary:

The key features to the Challenge 4 are:

Creating a Broken-Out Section view and Inserting the Model Dimensions.

CHALLENGE 6

This challenge examines your skills on the following:

* Saving a named view
* Creating a section view
* Measuring the perimeter of a surface

1. Opening a part document:

Select **File / Open**.

Browse to the Training Folder and open a part document named:
Block.sldprt.

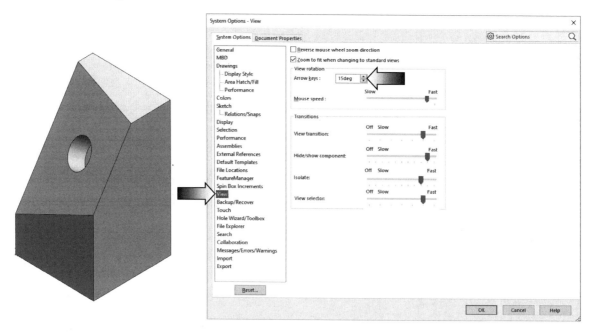

2. Changing the arrow key's angle:

The angle of the Arrow Key needs to change back to its default value for this challenge.

Click **Tools, Options, System Options, View**.

Change the angle of the Arrow Keys to **15deg**. and click **OK**.

3. Changing the view orientation:

Select the <u>face</u> indicated and press **Control+8** to rotate the model Normal to the screen.

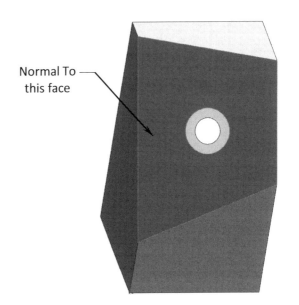

Normal To
this face

Hold the **Alt key** down and press the **Left Arrow** key <u>5 times</u>. The new orientation should look similar to the image shown at the bottom right of the page.

Press the **Spacebar** to access the Orientation dialog box.

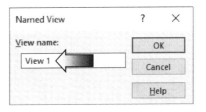

Click the <u>New View</u> button and enter **View 1** to save the new view.

Click **OK**.

The View 1 can now be accessed using the View Palette.
(Any drawing view can also be switched to the new View 1 by selecting its checkbox on the Annotation, More Views section.)

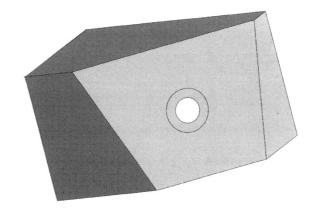

4. Making a drawing:

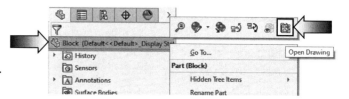

Right-click the name **Block** and select **Opening Drawing**.

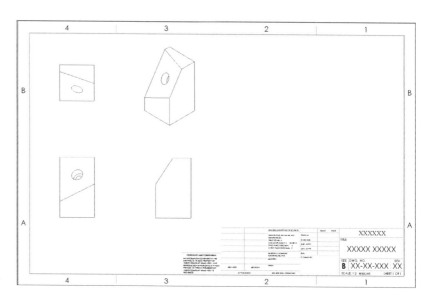

The 4 existing drawing views are for reference only. A couple of new drawing views need to be created.

Drag/drop the **Current View** from the **View Palette** as shown below.

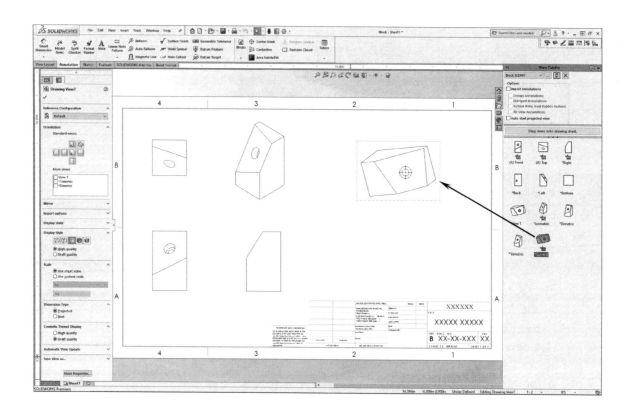

5. Adding a Projection Dimension:

Select the new drawing view and click **Projected** under the Dimension Type section, on the FeatureManager tree.

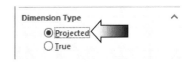

Add a <u>reference dimension</u> as shown in the image.

Enter the Projected dimension X

here _____ in.

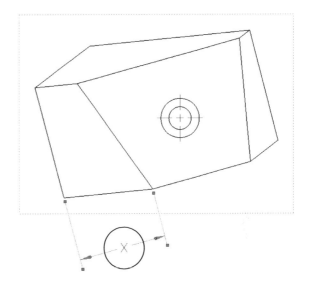

6. Creating a section view:

Switch to the **Sketch** tab and sketch a **Line** similar to the one shown in image below.

Add a **Concentric** relation between the <u>Midpoint</u> of the line and one of the circles.

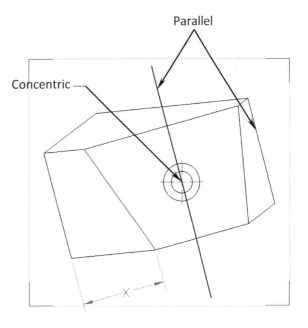

Add a **Parallel** relation between the line and the edge on the far right side.

Switch to the **View Layout** tab and click **Section View**.

7. Finding the surface Perimeter:

Place the section view on the left side of the main view.

The section arrows should be pointing to the right.

Switch to the **Evaluate** tab and click **Measure**.

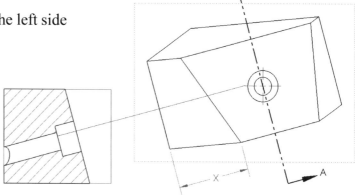

Select this face

Select the <u>face</u> indicated, locate the **Perimeter** value and enter it here:

_____ in.

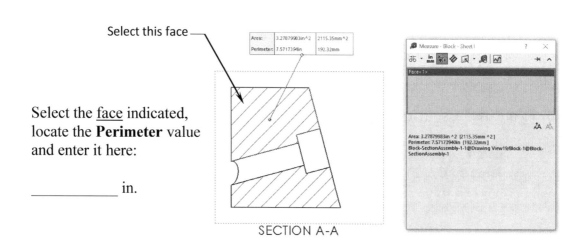

SECTION A-A

8. Saving your work:

Select **File, Save As**.

Enter **Challenge_6** for the file name.

Click **Save**.

Summary:

The key features to the Challenge 4 are:

Creating a Custom view and making a section from the view for measurements.

CHAPTER 2

CSWP – Advanced Mold Making

CSWP – Advanced Mold Making

Certified SOLIDWORKS Professional Advanced Mold Making.

The completion of the Certified SOLIDWORKS Professional Advanced Mold Making (CSWPA-MM) exam shows that you have successfully demonstrated your ability to use SOLIDWORKS Mold Tools functionality with Mold Making Industry knowledge.

Successful completion of this exam will demonstrate knowledge of how to correctly use these SOLIDWORKS tools.

Note: You must use at least SOLIDWORKS 2015 for this exam. Any use of a previous version will result in the inability to open some of the testing files.

Exam Length: 90 minutes

Minimum Passing grade: 75%

Re-test Policy: There is a minimum 30 day waiting period between every attempt of the CSWPA-MM exam. Also, a CSWPA-MM exam credit must be purchased for each exam attempt.

All candidates receive electronic certificates and a personal listing on the CSWP directory when they pass.

Exam features hands-on challenges in many of these areas of SOLIDWORKS Mold Tools functionality such as:

Parting Line creation, Parting Surface creation, Draft Analysis, Shut-off Surface creation, Imported part repair, Cavity Tool, and Parting Line Split Face.

CSWP – Advanced Mold Making
Interlock Surface

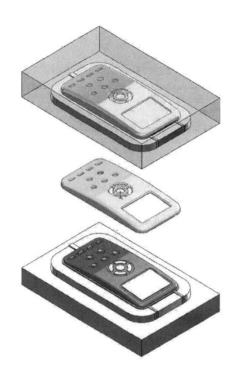

Dimensioning Standards: **ANSI**

Units: **INCHES** – 3 Decimals

Tools Needed:

Parting Lines	Shut-Off Surfaces	Parting Surfaces
Plane	Tooling Split	Mass Properties

1. Opening a part document:

Select **File / Open**.

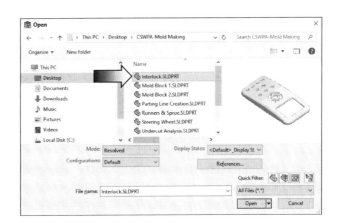

Browse to the Training Folder and open a part document named: **Interlock.sldprt**.

This exercise discusses the creation of the Interlock Surfaces.

2. Enabling the Mold Tools:

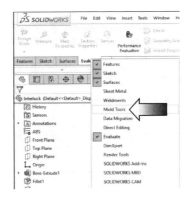

Right-click one of the tool tabs and enable the **Mold Tools** (arrow).

*Note: The **Default** should be the only configuration on the Configuration-Manager Tree; delete all others.*

Click the **Mold Tools** tab (arrow).

There are 4 sections in this tools tab to assist you with creating the mold:

    * Section 1: Preparing the surfaces or manually create Mold Tooling

    * Section 2: Analyzing and correcting the drafts or undercuts

    * Section 3: Adding drafts or scaling the model to accommodate the shrinkage

    * Section 4: Creating the mold and tooling split

3. Selecting the Parting Lines:

The Parting lines lie along the edge of the model, between the core (Red color) and the cavity (Green color). They are used to create the parting surfaces and to separate the two mold halves.

Click the **Parting Lines** command .

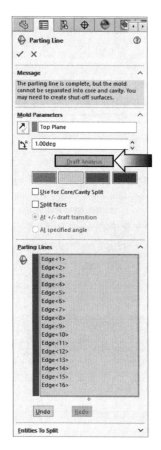

For Direction of Pull, select the **Top** plane from the Feature-Manager tree.

For Draft Angle, enter **1.00deg**.

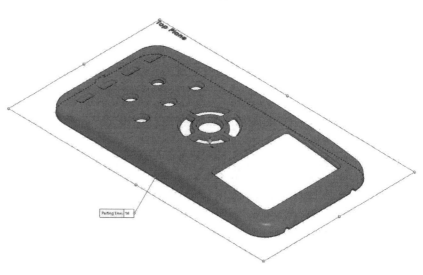

Click the **Draft Analysis** button (arrow).

The parting line for this model is selected automatically.
It is displayed in Blue color, around the perimeter of the part.

Parting Lines ⎯⎯

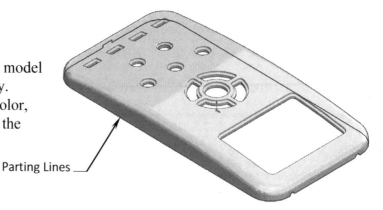

Click **OK**.

4. Creating the Shut-Off Surfaces:

A shut-off surface is used to close up all through holes in the plastic part by creating a surface patch along one of the edges of an opening that forms a continuous loop.

Click the **Shut-Off Surfaces** command 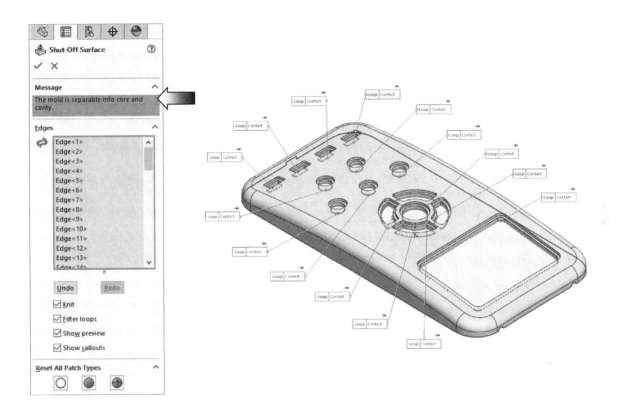.

Enable the **Knit** and **Show Callouts** checkboxes.

Leave the default Patch Type and **Contact** for all edges.

A <u>green message</u> appears on the top of the tree indicating: **"The mold is separable into core and cavity"**.

Click **OK**.

5. Creating the Parting Surfaces:

The Parting surfaces are used to split the mold cavity from the core.

Click the **Parting Surfaces** command .

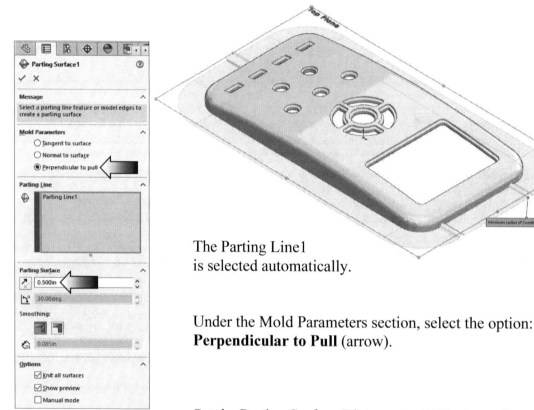

The Parting Line1
is selected automatically.

Under the Mold Parameters section, select the option:
Perpendicular to Pull (arrow).

Set the Parting Surface Distance to **.500in** (arrow).

Leave the Smoothing option at **Sharp**. (This option sets a distance between adjacent edges. A higher value creates a smoother transition between adjacent edges.)

Enable the **Knit All Surfaces** checkbox.

Click **OK**.

6. Adding a new plane:

To prevent the core and cavity blocks from shifting when opening and closing the mold, the interlock surface is added along the perimeter of parting surfaces, prior to inserting a tooling split in a plastic part.

The interlock surface usually has a draft angle between 3 and 5 degrees and its height is determined from the plane that we are going to create in this next step.

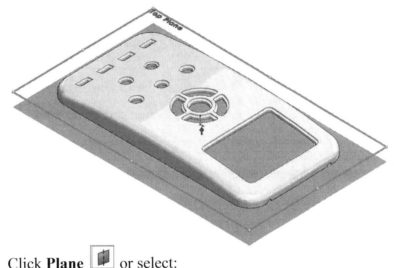

Click **Plane** or select:
Insert, Reference Geometry, Plane.

For First Reference, select the **Top** plane from the Feature-Manager tree.

Select the **Offset Distance** button and enter **.375in**.

Click the **Flip Offset** checkbox if needed to place the new plane <u>below</u> the Top.

Click **OK**.

7. Sketching the mold profile:

Open a **new sketch** on Plane1.

Sketch a **Corner Rectangle** as shown.

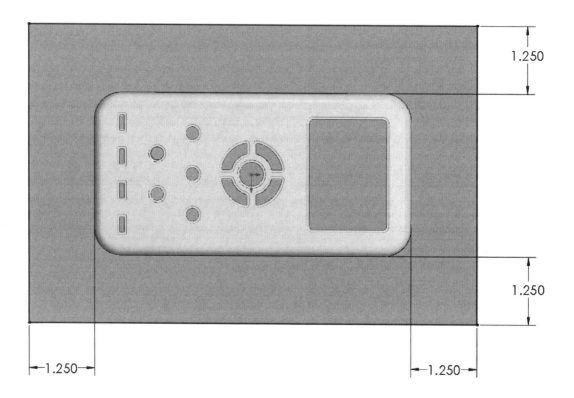

Add the location **dimensions** to fully define the sketch.

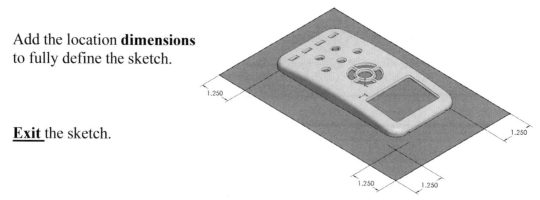

Exit the sketch.

(The sketch must be selected prior to Tooling Split.)

8. Performing a Tooling Split:

Tooling Split is used to separate the core and cavity.

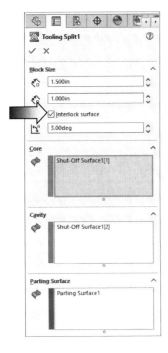

Switch back to the Mold Tools tab and click **Tooling Split** .

For Block Size, enter:

1.500in for the upper block

1.000in.for the Lower block.

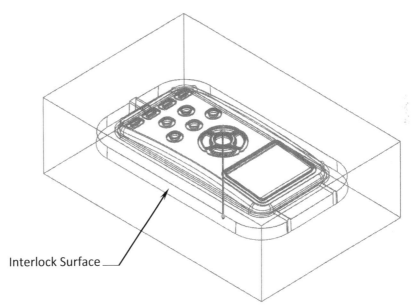

Interlock Surface

Enable the **Interlock Surface** checkbox.

Enter **3deg**. for the draft angle on the interlock surfaces.

The preview graphics show the interlock surfaces is being created.

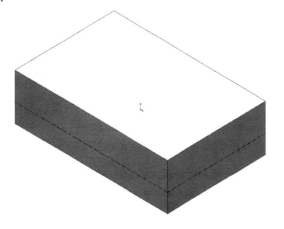

Click **OK**.

9. Separating the solid bodies:

The next step is to separate the 2 mold blocks to inspect them.

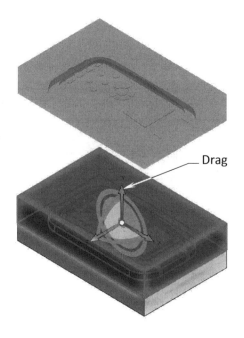

Select: **Insert, Features, Move/Copy** (arrow).

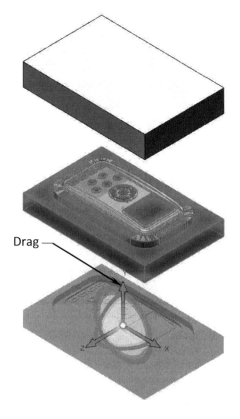

Drag

Select the Upper mold block and drag the **vertical arrowhead** (Y direction) upwards, approximately **7.300 inches**.

Click **OK**.

Repeat step 9 and move the Lower mold block downwards, approximately the same distance.

Click **OK**.

Drag

10. Assigning materials:

Expand the **Solid Bodies** folder. (The Cavity and Core bodies have been renamed for clarity only.)

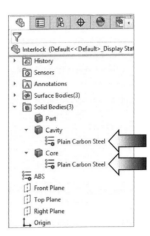

Expand the **Cavity** body. Right-click the Material option and select: **Plain Carbon Steel**.

Expand the **Core** body. Right-click the Material option and once again, select: **Plain Carbon Steel**.

11. Calculating the Mass:

Switch to the **Evaluate** tab and click **Mass Properties** ⚖.

Select only the **Core** body (arrow) to calculate the mass.

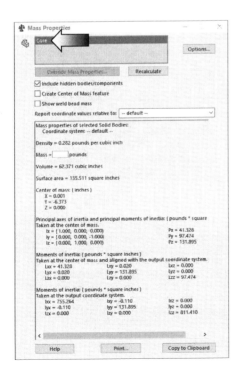

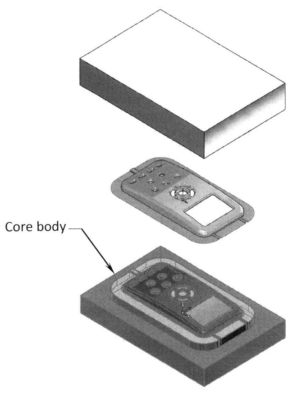

Core body

Enter the mass of the **Core** body here: _____ lbs.

12. Saving your work:

Select **File, Save**.

<u>Replace</u> the previous document with your completed one.

Close all documents.

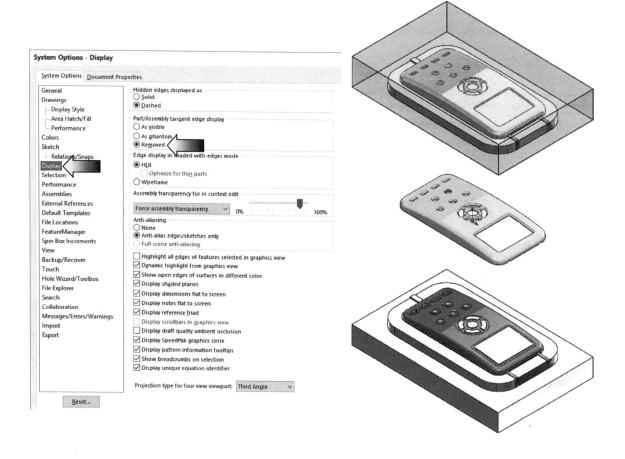

The image shown here was created using the following settings:

    * **Part/Assembly Tangent Edge Display**: **Removed** (arrow).

    * **Realview Graphics** enabled.

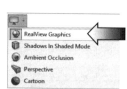

CSWP – Advanced Mold Making
Parting Line Creation

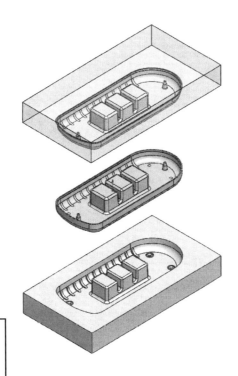

Dimensioning Standards: **ANSI**

Units: **INCHES** – 3 Decimals

Tools Needed:

 Parting Lines

 Corner Rectangle

Shut-Off Surfaces

Tooling Split

 Parting Surfaces

 Mass Properties

1. Opening a part document:

Select **File / Open**.

Browse to the Training Folder and open the part document named: **Parting Line-Creation.sldprt**.

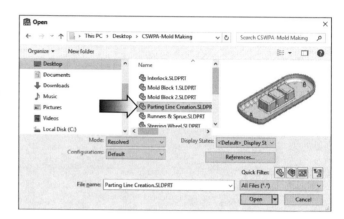

2. Enabling the Mold Tools:

Right-click one of the tool tabs and enable the **Mold Tools** (arrow).

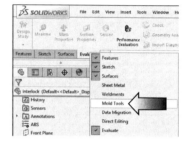

3. Creating the Parting Lines:

Click **Parting Lines** on the Mold Tools tab.

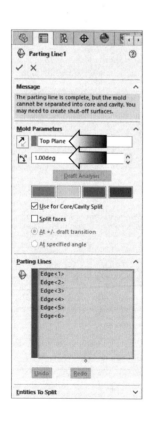

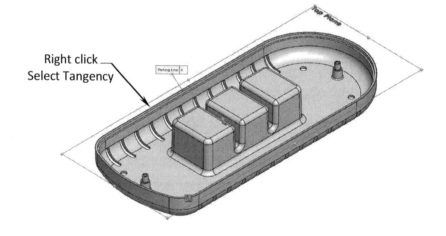

Right click
Select Tangency

For Direction of Pull, select the **Top** plane from the tree.

For Draft Angle, enter **1deg**.

For Parting Lines, right-click one of the outer edges and pick: **Select Tangency**.

Click **OK**.

4. Creating the Shut-Off Surfaces:

Shut-off surfaces close up the through holes in a part. Once the shut-off surfaces are created, SOLIDWORKS places them into the Cavity Surface Bodies and Core Surface Bodies folders.

Select the **Shut-Off Surfaces** command .

The edges of all the openings in the part are selected automatically. Verify that there are no redundant edges (red color). Select the edges manually if needed.

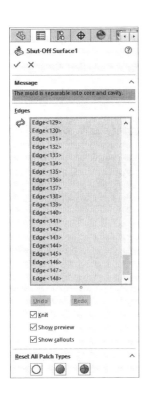

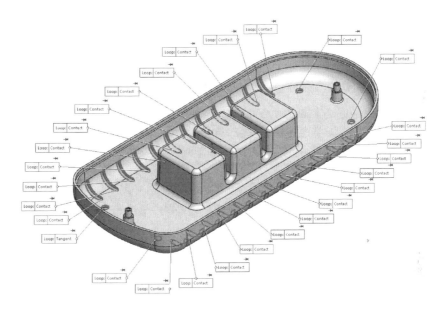

Enable the **Knit** and **Show Callouts** checkboxes.

A green message on the upper left side of the Properties tree indicates that the mold is separable into Core and Cavity.

Click **OK**.

5. Creating the Parting Surfaces:

The Parting Surfaces are created after Parting Lines and Shut-Off Surfaces. The Parting Surfaces split the mold cavity from the core.

Select the **Parting Surfaces** command .

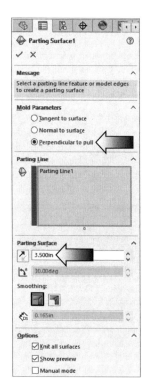

For Mold Parameters, select **Perpendicular to Pull** (arrow).

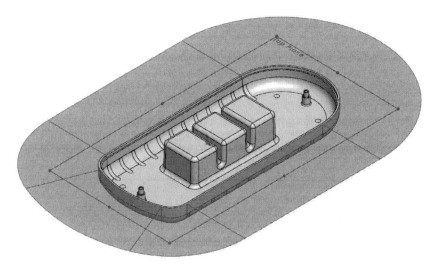

For Parting Surface Distance, enter **3.500in**.

Keep the Smoothing at the default **Sharp**.

Enable the **Knit All Surfaces** checkbox.

Click **OK**.

6. Sketching the mold block profile:

Select the <u>Top</u> plane and open a **new sketch**.

Sketch a **Corner Rectangle** just a little larger than the part.

Add the dimensions shown to fully define the sketch.

The dimensions are measured from the origin to the lines.

(The width dimensions are symmetrical, not the height.)

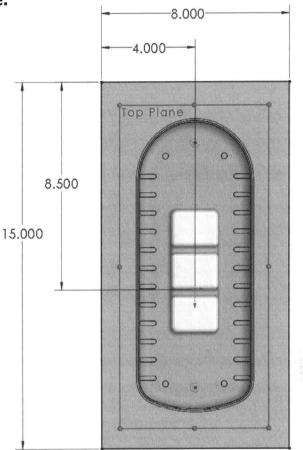

Exit the sketch.

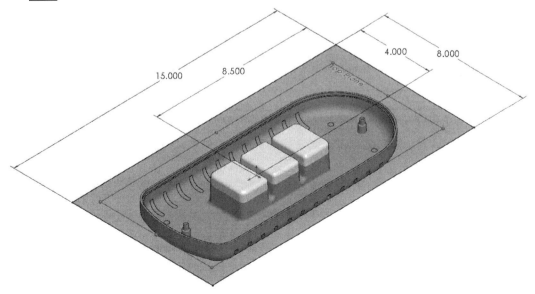

7. Inserting a Tooling Split:

Select **Sketch1** either from the graphics area or from the FeatureManager tree.

Click the **Tooling Split** command .

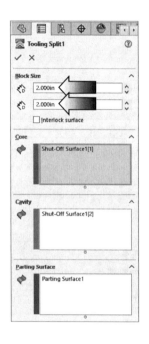

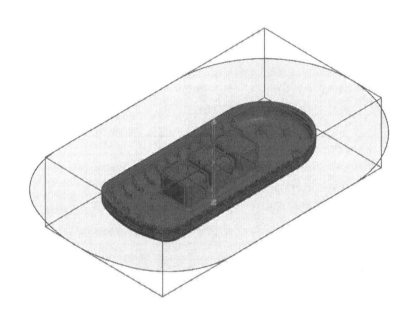

For Block Size, enter the following:

      * **2.000in** upper block

      * **2.000in** lower block

Clear the Interlock Surface checkbox.

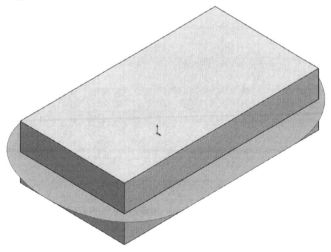

Click **OK**.

8. Separating the mold blocks:

The next step is to separate the 2 mold blocks to inspect them.

Select: **Insert, Features, Move/Copy** (arrow).

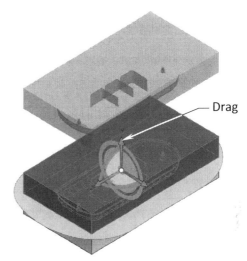

Select the Upper mold block and drag the **vertical arrowhead** (Y direction) upwards, approximately 8.250 inches.

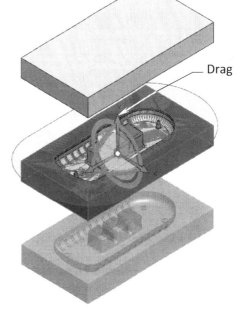

Repeat step 8 and move the Lower mold block downwards, about the same distance as the first one.

Click **OK**.

9. Assigning materials:

Expand the **Solid Bodies** folder. (The Cavity and Core bodies have been renamed for clarity only.)

Expand the **Cavity** body. Right-click the Material option and select: **Plain Carbon Steel**.

Expand the **Core** body. Right-click the Material option and also select: **Plain Carbon Steel**.

10. Calculating the Mass:

Switch to the **Evaluate** tab and click **Mass Properties** .

Select only the **Cavity** body (arrow) to calculate the mass.

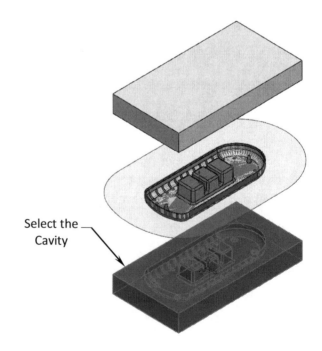

Select the Cavity

Enter the mass of the **Cavity** body here: _____ lbs.

11. Saving your work:

Select **File, Save**.

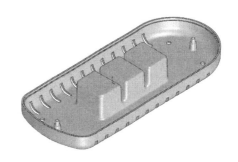

Overwrite the previous document with your completed one.

Close all documents.

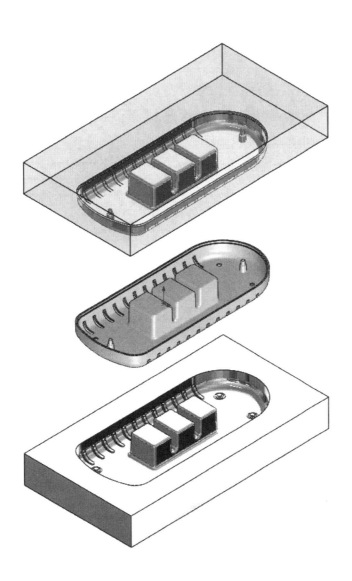

CSWP – Advanced Mold Making
Runners & Sprue

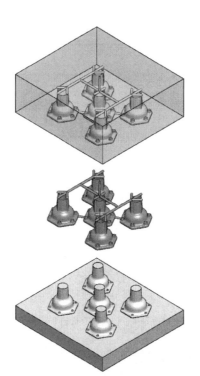

Dimensioning Standards: **ANSI**

Units: **INCHES** – 3 Decimals

Tools Needed:

Parting Lines	Shut-Off Surfaces	Parting Surfaces
Scale	Measure	Mass Properties

1. Opening a part document:

Select **File / Open**.

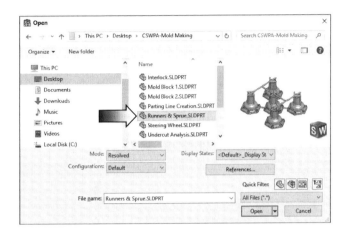

Browse to the Training Folder and open a part document named:
Runners & Sprue.sldprt.

This exercise discusses one of the special techniques to create a tooling split with Runners and Sprue.

2. Creating the Parting Lines:

Change to the **Mold Tools** tab.

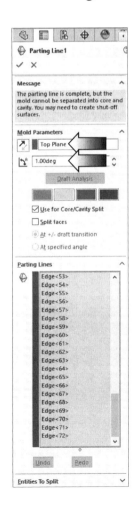

Select the **Parting Lines** command 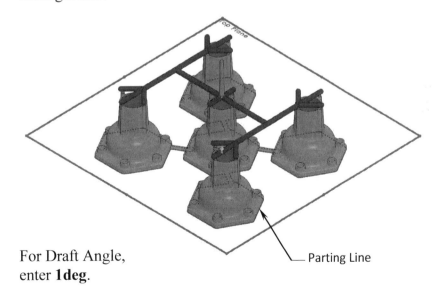.

For Mold Parameters, select the **Top** plane from the Feature-Manager tree.

Parting Line

For Draft Angle, enter **1deg**.

Click **OK**. The parting Line1 is created along the perimeter at the bottom of the part.

3. Creating the Shut-Off Surfaces:

There are several through openings in this part. They need to be closed off.

Select the **Shut-Off Surfaces** command 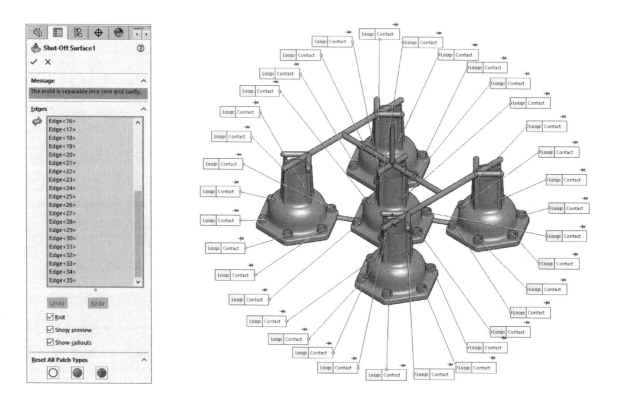.

The edges that form the through openings are selected and Shut-Off Surfaces are created to close them off.

Enable the **Knit** and **Show Callouts** checkboxes.

A green-color message appears on the upper left side of the PropertiesManager tree indicating the mold is separable into Core and Cavity.

Click **OK**.

4. Creating the Parting Surfaces:

Click the **Parting Surfaces** command 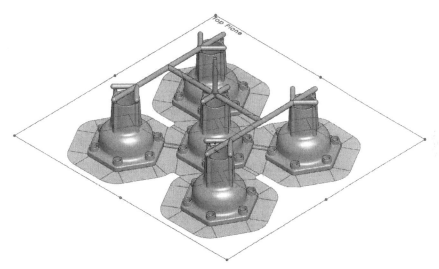.

For Mold Parameters, select the **Perpendicular to Pull** option (arrow).

The Parting Line1 is selected automatically.

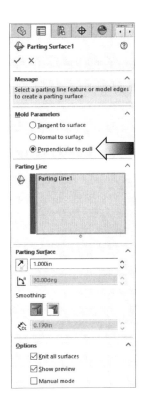

For Parting Surface Distance, enter **1.000in**.

Keep the Smoothing option at its default **Sharp**.

Enable the **Knit All Surfaces** and **Show Preview** checkboxes.

The preview graphics show the Parting Surface being created.

Click **OK**.

5. Sketching the preliminary shape of the mold block:

Due to the unique geometry of the parting surfaces, a couple of extra steps must be added in order to split the mold later on.

Select the <u>Top</u> plane and open a **new sketch**.

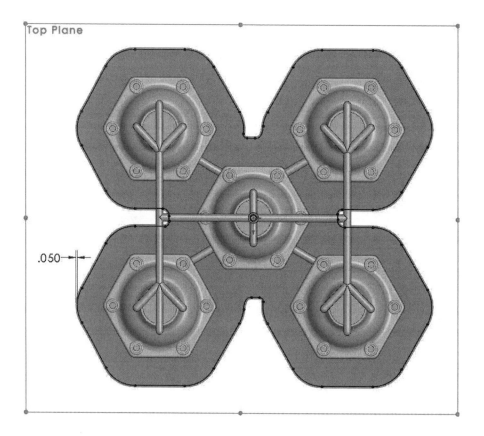

Right-click one of the outer edges of the parting surface and pick:
Select Tangency.

Click **Offset Entities** and create an Offset Distance of **.050"** smaller (inside).

<u>**Exit**</u> the Sketch.

6. Inserting a Tooling Split:

Switch to the **Mold Tools** tab.

Select the sketch from the last step and click **Tooling Split** 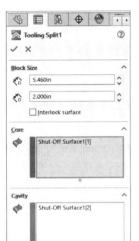.

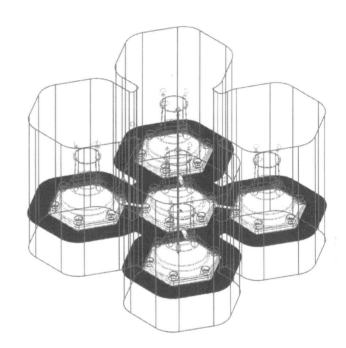

For Block Size, enter the following:

    * **5.460in** upper block

    * **2.000in** lower block

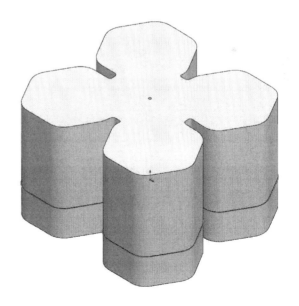

Clear the **Interlock Surface** option.

Click **OK**.

The next step is to finalize the shape of the 2 mold blocks.

7. Finalizing the sketch of the lower mold block:

This step finalizes the shape of the mold block by extruding a rectangle to square off the 4 sides.

Select the Top plane and open a **new sketch**.

Create an Offset of **.050"** (smaller) from the edges of the parting surfaces.

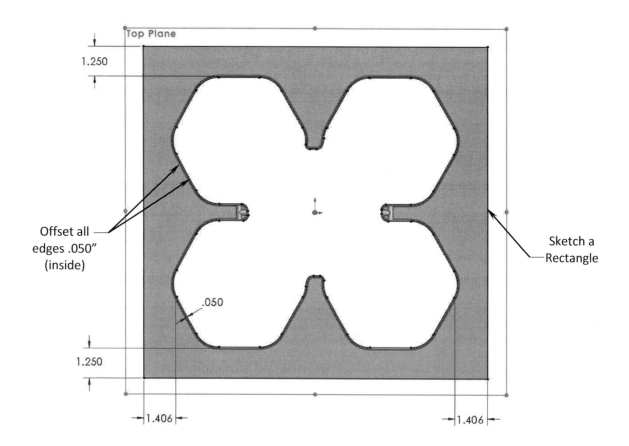

Offset all edges .050" (inside)

Sketch a Rectangle

Sketch a **Rectangle** around the perimeter of the sketch outline and add the location **dimensions** shown to fully define the sketch.

8. Extruding the lower mold block:

Switch to the **Features** tool tab.

Click **Extruded Boss/Base** .

For Direction 1, select **Up To Surface** (arrow).

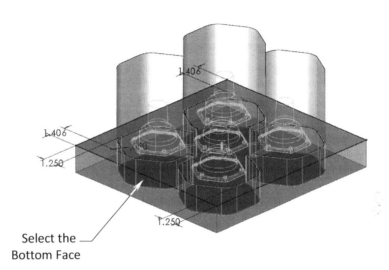

Select the
Bottom Face

Rotate the model and select the **Bottom Face** as noted.

Expand the **Feature Scope** section
and click the **Selected Bodies**
option (arrow).

Select the **Core** body either
in the graphics area or from
the FeatureManager tree.
This extrude feature will
blend to the Core body.

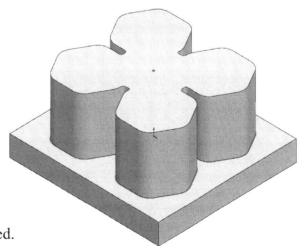

Click **OK**. The Core block is created.

9. Converting a sketch:

The next step is to create the Cavity block.

Expand the **Boss-Extrude1** feature and <u>show</u> the sketch under it (arrow).

Open a **new sketch** on the <u>Top</u> plane.

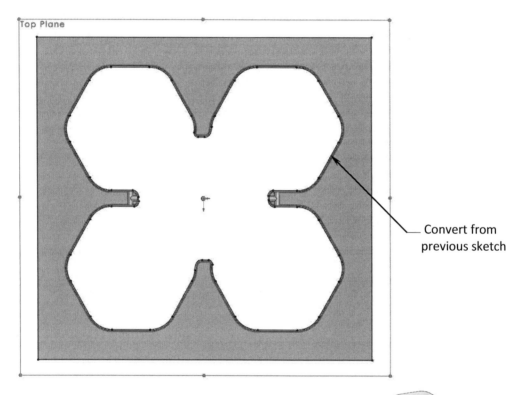

Convert from previous sketch

Select the **Sketch3** created from the last step and press **Convert Entities** 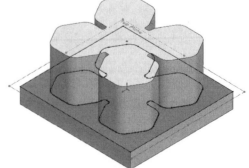.

The selected entities from the previous sketch are converted to a new sketch.

10. Extruding the upper mold block:

Switch to the **Features** tab and click **Extruded Boss/Base** .

For Direction 1, select **Up To Surface** (arrow).

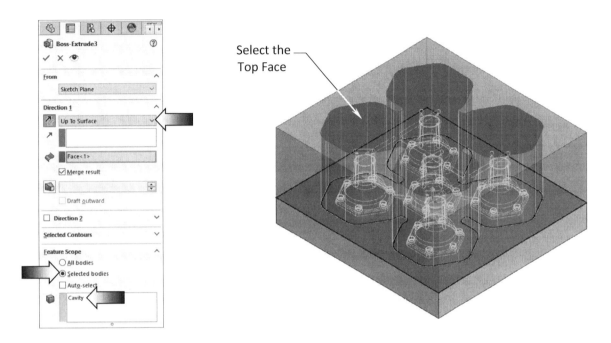

Select the Top Face

Press **Control+7** to change to the **Isometric** view.

Select the **Top Face** of the part as noted.

Expand the **Feature Scope** section and click the **Selected Bodies** option (arrow).

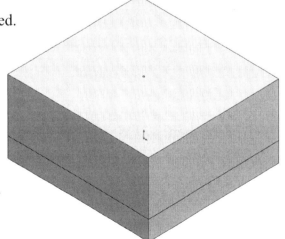

Select the **Cavity** body from the graphics area or from the Feature tree. This extrude feature will blend to the Cavity body.

Click **OK**. Cavity block is created.

11. Moving the Cavity mold block:

The height of the part is exactly 3.00 inches. We will move the cavity block upwards to simulate the opening of the mold.

Select **Insert, Features, Move/Copy**.

Select the **Cavity** block and enter **3.000in** in the **Delta Y** field.

Click **OK**.

The Cavity block moves up 3 inches.

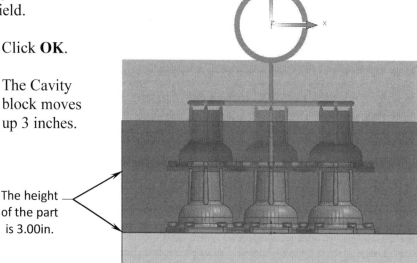

The height of the part is 3.00in.

12. Measuring the distance:

Change to the **Evaluate** tab and click the **Measure** command .

Change to the **Front** orientation or press **Control + 1**.

Using the Measure tool, select the <u>Top</u> and <u>Bottom</u> edges as shown.

Enter the **Normal Distance (X)** that represents the maximum mold opening here:

_____ inches.

13. Copying the plastic part:

The next step is to make a copy of the plastic part and use it to subtract the Cavity block.

Select **Insert, Features, Move/Copy**.

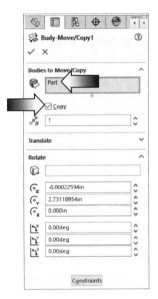

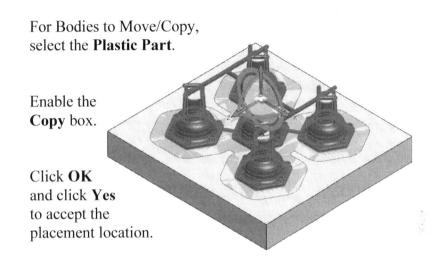

For Bodies to Move/Copy, select the **Plastic Part**.

Enable the **Copy** box.

Click **OK** and click **Yes** to accept the placement location.

14. Combine-Subtracting the solid bodies:

Select **Insert, Features, Combine**.

Select the **Subtract** option (arrow).

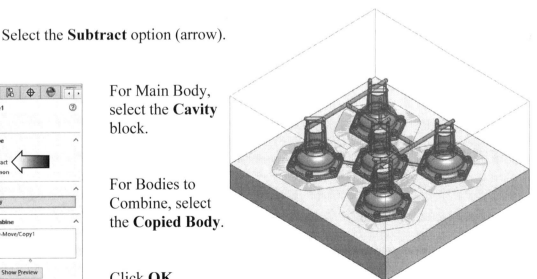

For Main Body, select the **Cavity** block.

For Bodies to Combine, select the **Copied Body**.

Click **OK**.

15. Applying Shrink Rate (Scale):

Switch to the **Mold Tools** tab and click **Scale** .

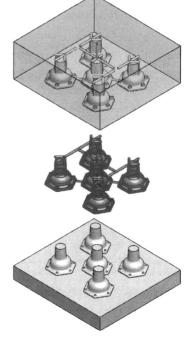

Select the **Part** from the Solid Bodies folder. (The Exploded view is for clarity only.)

Enter the scale of **1.05** (5% larger).

For Scale About, use the default **Centroid** option.

Click **OK**.

16. Calculating the total mass:

Switch to the **Evaluate** tab and click **Mass properties** .

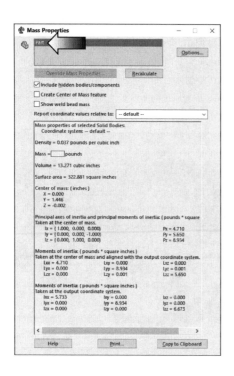

Select only the **Plastic Part** to calculate the mass. (Its material has already been set to ABS.)

Enter the Mass of the Part here: _____ lbs.

17. Saving your work:

Select **File, Save**.

Use the same file name to save the document and allow to overwrite when prompted.

Close all documents.

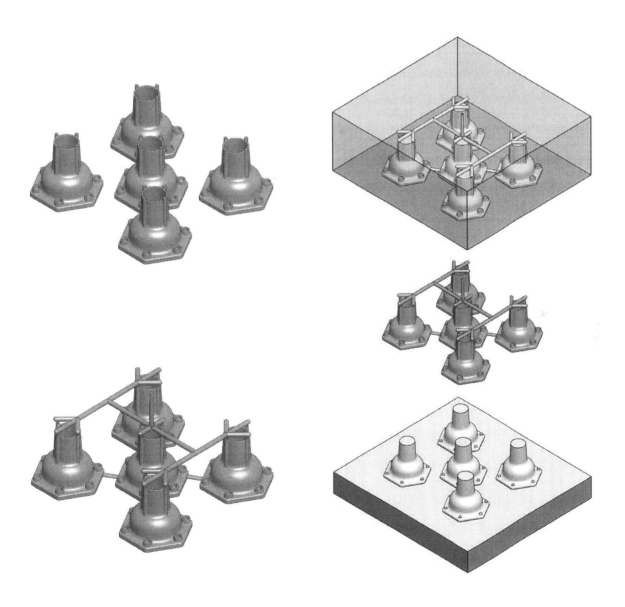

CSWP – Advanced Mold Making
Shut-Off Surface Creation

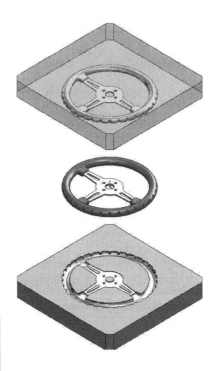

Dimensioning Standards: **ANSI**

Units: **INCHES** – 3 Decimals

Tools Needed:

 Split Line   Parting Lines   Shut-Off Surfaces

 Measure   Parting Surface   Tooling Split

1. Opening a part document:

Select **File / Open**.

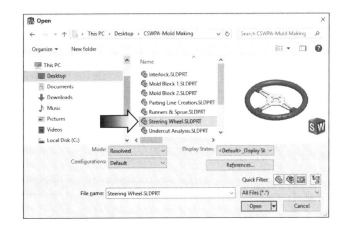

Browse to the Training Folder and open the part document named:
Steering Wheel.sldprt.

This exercise discusses the creation of the Shut-Off Surfaces.

2. Creating a Split Line:

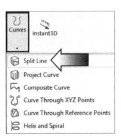

The Split Line command projects a sketch entity to surfaces, or curved or planar faces. It divides a selected face into multiple separate faces.

Select the <u>Front</u> plane and open a **new sketch**.

Sketch a horizontal **Line** across the model as shown below.

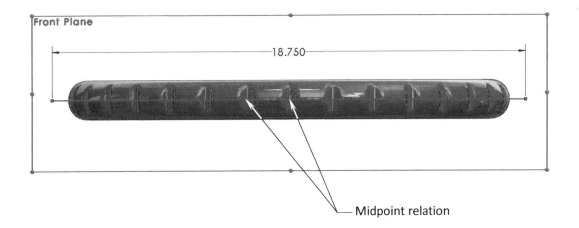

Add a **Midpoint** relation and an overall length **Dimension**.

Switch to the **Features** tool tab and select: **Curves, Split Line** .

For Type of Split, select **Projection** (arrow).

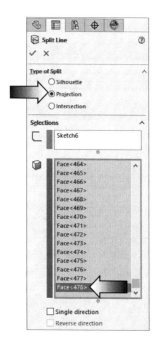

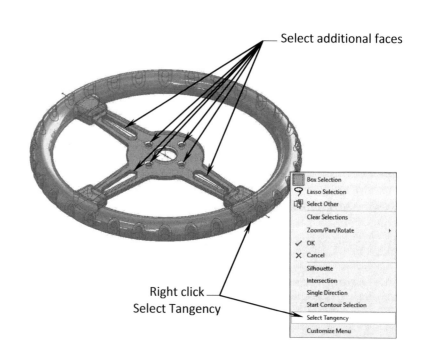

Select additional faces

Right click
Select Tangency

For Faces to Split, right-click the face of the wheel and pick: **Select Tangency**.

Select the additional faces as indicated. Use the Select Tangency option to select multiple faces more quickly.

Zoom in as needed to correctly select all the faces around the wheel.

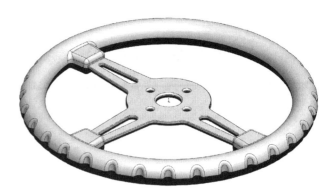

There should be a total of **478 faces** to split (arrow).

Click **OK**.

3. Creating the Parting Lines:

Select the **Parting Lines** command 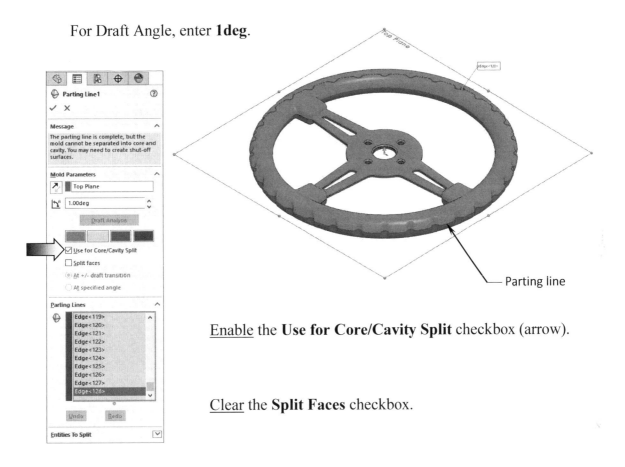 on the Mold Tools tab.

For Mold Parameters, select the **Top** plane from the FeatureManager tree.

For Draft Angle, enter **1deg**.

Parting line

Enable the **Use for Core/Cavity Split** checkbox (arrow).

Clear the **Split Faces** checkbox.

The Parting Line is selected automatically, but it lies only along the outer perimeter of the model.

The **green color** faces represent the Positive side of the mold (top), and they will be used to create the upper half of the mold.
The **red color** is the Negative side (bottom).

Click **OK**.

4. Creating the Shut-Off Surfaces:

Select the **Shut-Off Surfaces** command on the Mold Tools tab.

The edges in the model that form a closed boundary are selected. Ensure that the edges of the 5 holes and the 3 slots are also selected; if not, select them manually.

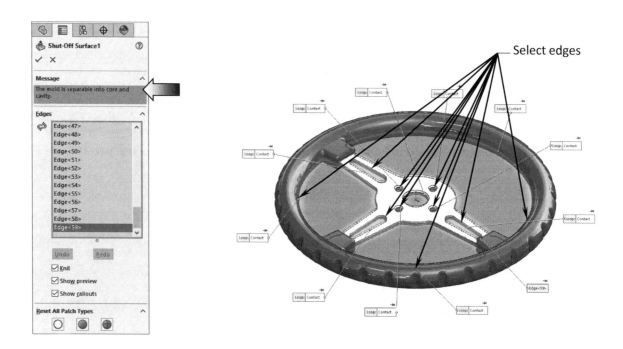

The preview graphic shows the Shutoff surfaces are being created, and the message on the Property tree changes to green color (arrow) indicating the mold is separable into core and cavity.

Enable the **Knit**, **Show Preview**, and **Show Callout** checkboxes.

There should be a total of **59 edges**, or **11** shut-off surfaces.

Click **OK**.

5. Measuring the surface area:

Switch to the **Evaluate** tab and click **Measure** .

Select the **Show XYZ Measurements** option (arrow).

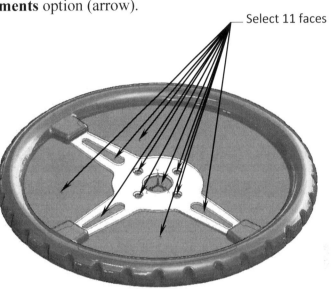

Select 11 faces

Select all **11 Shut-Off Surfaces**
created in the previous step.

Enter the **Total Surface Area** of the Shut-Off Surfaces here: _____ in^2

(This portion of the exam stops at this point after the creation of the Shut-Off Surfaces is completed; but for practice purposes, we will go ahead and continue with the last few steps and create the 2 mold blocks for the Steering Wheel.)

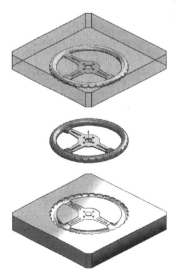

6. Creating the Parting Surfaces:

Change back to the **Mold Tools** tab.

Select the **Parting Surfaces** command .

For Mold Parameters, select the **Perpendicular to Pull** option (arrow).

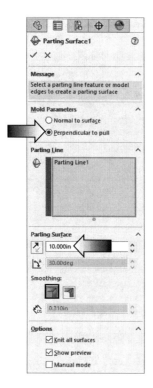

The Parting Line1 is selected automatically.

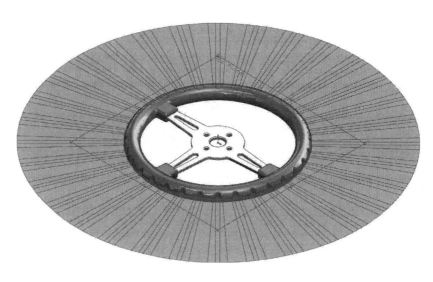

For Parting Surface Distance, enter **10.000in**.

Keep the Smoothing at **Sharp** and enable the option **Knit All Surfaces**.

Click **OK**.

7. Inserting a Tooling Split:

Select the Top plane and open a **new sketch**.

Sketch a **Center-Rectangle** that is centered on the origin.

Add an **Equal** relation between 1 horizontal line and 1 vertical line.

Add an overall height dimension to one of the lines to fully define the sketch.

25.000

Exit the sketch. Switch to the **Mold Tools** tab and click **Tooling Split**.

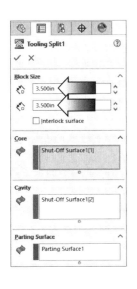

Enter the following for Block Size:

* **3.500in** upper block
* **3.500in** lower block

Clear the Interlock Surfaces checkbox.

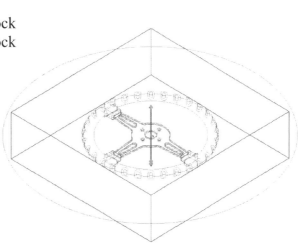

Click **OK**.

8. Adding chamfers:

Switch to the **Features** tab.

Select the **Chamfer** command
(under Fillet).

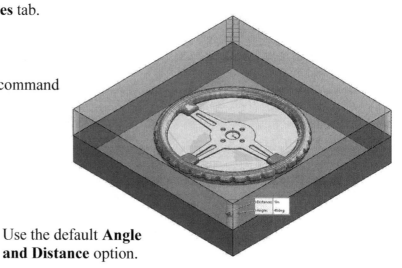

Use the default **Angle
and Distance** option.

For Chamfer Depth, enter **1.000in**.

For Chamfer Angle, enter **45.00deg**.

Select the **4 vertical edges** of the upper block.

Click **OK**.

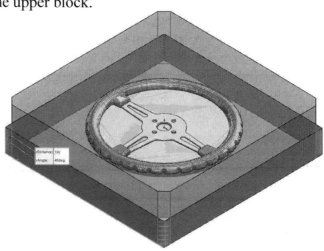

Since the 2 blocks are
2 different bodies, the
chamfer feature has to be
added as 2 different features.

Repeat the step and add the same chamfer to the lower block.

9. Creating an exploded view:

Exploded views in a Multibody part are used for viewing purposes. Editing the solid bodies is prohibited when the exploded view is active. Exploded Views are stored in the Configuration tree, under the Default configuration.

Select **Insert, Exploded View**.

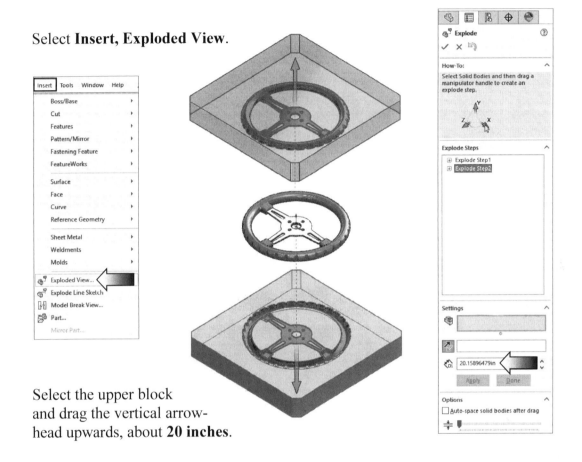

Select the upper block and drag the vertical arrow-head upwards, about **20 inches**.

Select the lower block; drag the vertical arrowhead downwards, about 20 inches.

10. Saving your work:

Select **File, Save**.

Use the same file name to save and allow overwriting when prompted.

Close all documents.

CSWP – Advanced Mold Making
Undercut Analysis

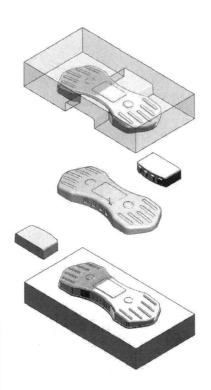

Dimensioning Standards: **ANSI**

Units: **INCHES** – 3 Decimals

Tools Needed:

 Parting Lines   Shut-Off Surfaces   Parting Surfaces

 Undercut Analysis   Tooling Split   Slide Core

1. Opening a part document:

Select **File / Open**.

Browse to the Training Folder and open a part document named: **Undercut Analysis.sldprt**.

This exercise discusses the use of the Undercut Analysis and the creation of a side core.

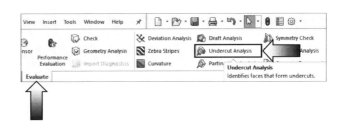

2. Performing an Undercut Analysis:

Undercut Analysis tool helps identify and visualize trapped areas on molded parts that would prevent the part from ejecting from the mold.

Switch to the **Evaluate** tab and select the **Undercut Analysis** command.

For Direction of Pull, select the **Top** plane from the FeatureManager tree.

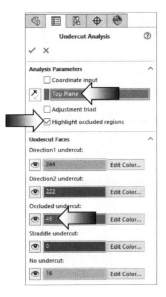

Enable the **Highlight Occluded Regions** checkbox.

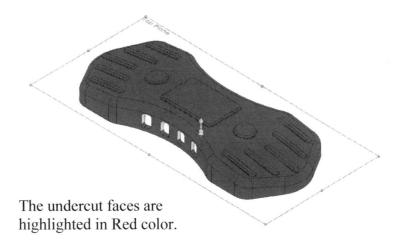

The undercut faces are highlighted in Red color.

How many Occluded Undercut faces will require a side core? (Check one)

☐ 12   ☐ 24   ☐ 36   ☐ 48

3. Creating the Parting Lines:

Change to the **Mold Tools** tab.

Click the **Parting Lines** command 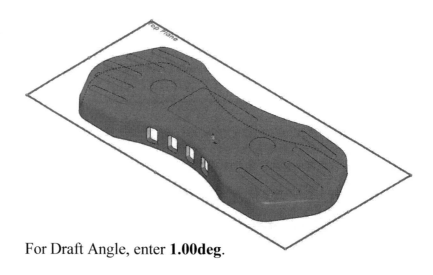.

For Mold Parameters, select the **Top** plane from the FeatureManager tree (arrow).

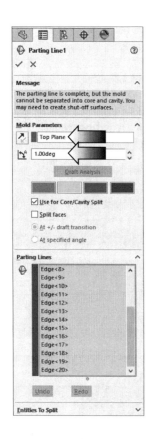

For Draft Angle, enter **1.00deg**.

Enable the **Use for Core/Cavity Split** checkbox.

The edges along the perimeter of the part are selected and will be used as the Parting lines for this model.

The **green color** faces will be used to create the core half of the mold and the **red color** faces are used to create the core half.

Click **OK**.

4. Creating the Shut-Off Surfaces:

There are several openings on the left and right sides of the model; they will need to be closed off.

Select the **Shut-Off Surfaces** command .

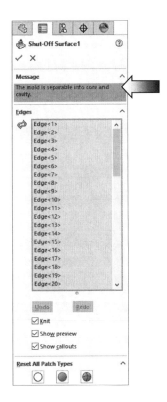

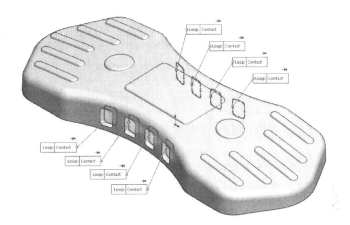

The inner edges of the rectangular holes are selected, and new shut-off surfaces are created to close them off.

If there are redundant edges (red color), use manual method to select all inner edges of the 8 holes, if needed.

Enable the **Knit** and **Show Callout** checkboxes.

A **green message** appears on the upper left side of the Properties tree indicating the mold is separable into Core and Cavity.

Click **OK**.

5. Creating the Parting Surfaces:

The Parting Surfaces are used to split the mold into core and cavity blocks.

Select the **Parting Surfaces** command .

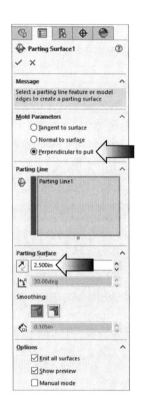

For Mold Parameters, select **Perpendicular to Pull** (arrow).

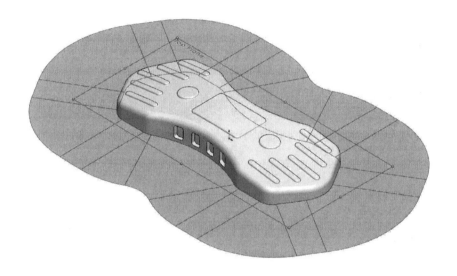

For Parting Surface Distance, enter **2.500in**.

Keep the Smoothing option at the default **Sharp**.

Enable the **Knit all Surfaces** and **Show preview** checkboxes.
The Preview graphics show the parting surfaces being created.

Click **OK**.

6. Inserting a Tooling Split:

Select the Top plane and open a **new sketch**.

Sketch a **Center Rectangle** and add the dimensions shown.

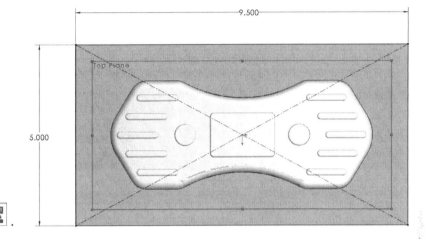

Exit the Sketch.

Select the sketch
and click:
Tooling Split 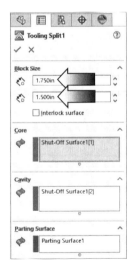.

Enter the following for Block Size:

* **1.750in** upper block

* **1.500in** lower block

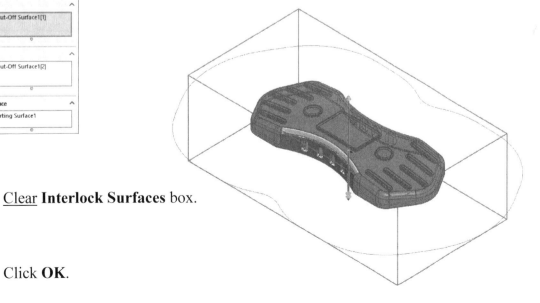

Clear **Interlock Surfaces** box.

Click **OK**.

7. Adding the 1st slide core:

The undercuts on the 2 sides of the model must be captured with a couple of slide cores.

Select the <u>Front</u> plane and open a **new sketch**.

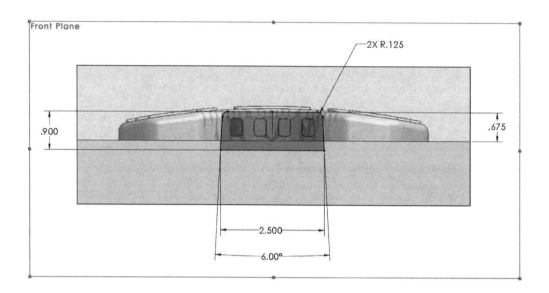

Sketch the profile shown above. Add dimensions and relations needed to fully define the sketch.

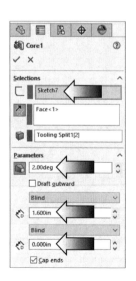

Exit the sketch.

Select the sketch of the core and click the **Core** command 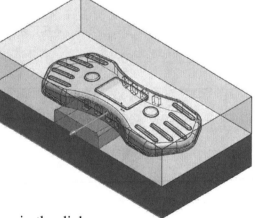.

Enter the **Parameters** shown in the dialog. Also enable the **Cap Ends** checkbox.

Click **OK**.

8. Adding the 2nd slide core:

Open a **new sketch** on the <u>back side</u> of the part.

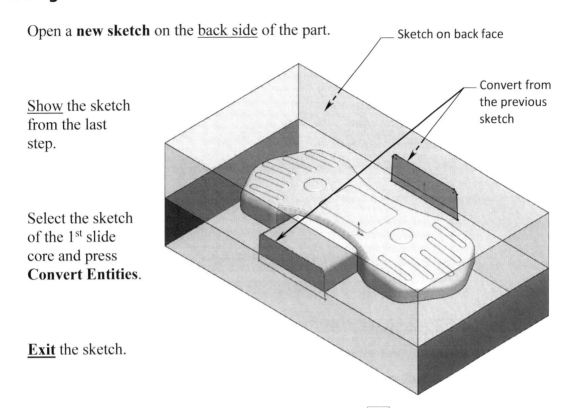

Sketch on back face

Convert from
the previous
sketch

<u>Show</u> the sketch
from the last
step.

Select the sketch
of the 1st slide
core and press
Convert Entities.

Exit the sketch.

Select the sketch of the 2nd slide core and click **Core**.

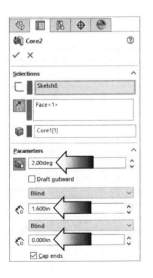

Enter the **Parameters**
shown on the left side.

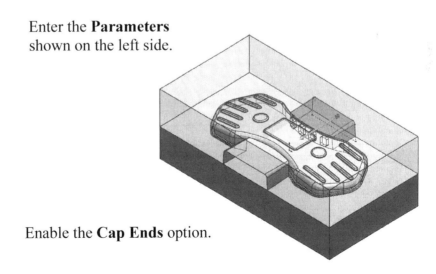

Enable the **Cap Ends** option.

Click **OK**.

9. Assigning material:

Expand the **Solid Bodies** and the **Core Bodies** folders.

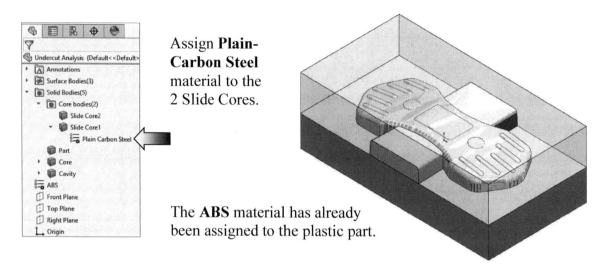

Assign **Plain-Carbon Steel** material to the 2 Slide Cores.

The **ABS** material has already been assigned to the plastic part.

10. Calculating the mass:

Switch to the **Evaluate** tab and click **Mass Properties** .

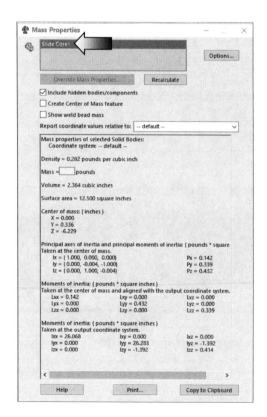

Select <u>one</u> of the **Slide Cores** either from the graphics area from the FeatureManager tree.

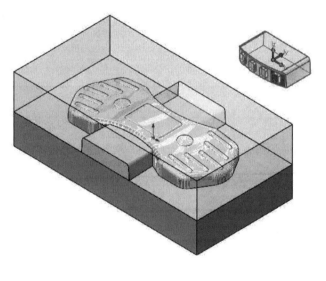

Enter the mass of the **Core** here: _____ lbs.

11. Saving your work:

Select **File, Save**.

Use the same file name and allow to overwrite when prompted.

Close all documents.

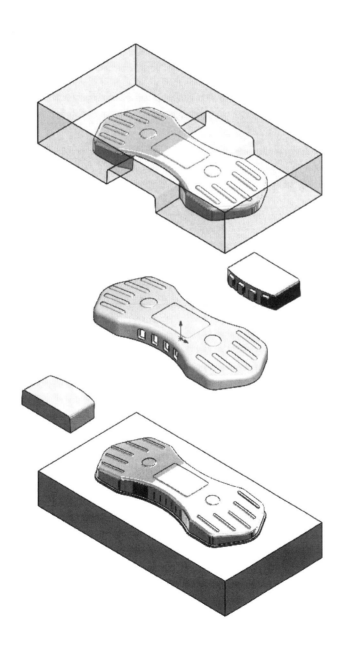

CSWP – Advanced Mold Making
Virtual Part Creation

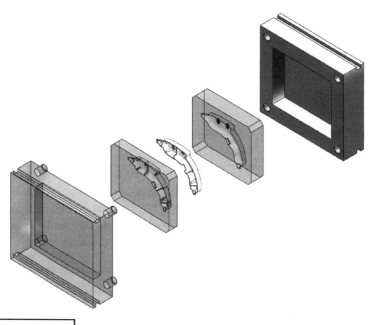

Dimensioning Standards: **ANSI**

Units: **INCHES** – 3 Decimals

Tools Needed:

 Insert New Part   Insert part   Move/Copy

 Intersect   Combine   Mass Properties

1. Opening a part document:

Select **File / Open**.

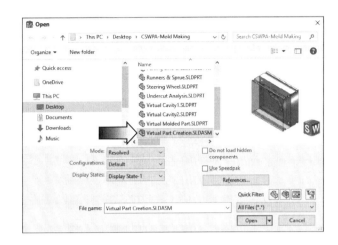

Browse to the Training Folder and open an assembly document named:
Virtual Part Creation.sldprt.

This exercise discusses one of the methods to create a Virtual Part in an assembly.

2. Hiding components:

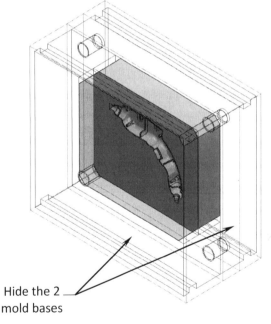

We will not need the 2 mold bases for a few moments, so let us hide them.

Click one of the mold bases and select **Hide** ⬚. Also hide the 2ⁿᵈ base.

The 2 components were made transparent for clarity only.

Hide the 2 mold bases

The exploded view was created only to show the Core and Cavity components. They will be used to create the plastic part and saved as a Virtual Component.

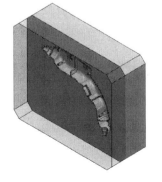

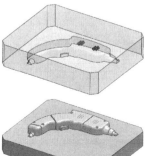

3. Inserting a New Part:

Switch to the **Assembly** tab.

Select the **New Part** option under **Insert Components** (arrow).

Select the **Front** plane of the assembly to reference the new part.

A new component is created and appears on the Feature tree, in Blue color (arrow).

Select Front plane from Feature tree

Click off the **Edit Component** command .

4. Saving the assembly:

The assembly document needs to be saved before any modifications can be done to the new component.

Select **File, Save**.

Click **Save All** and overwrite the previous document if prompted.

5. Working in the Part level:

For this particular part, we will create it in the part-level instead.

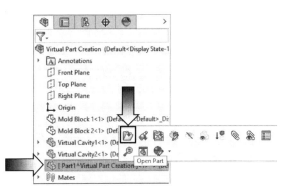

Select the **Part1** from the Feature tree and click **Open Part** (arrow).

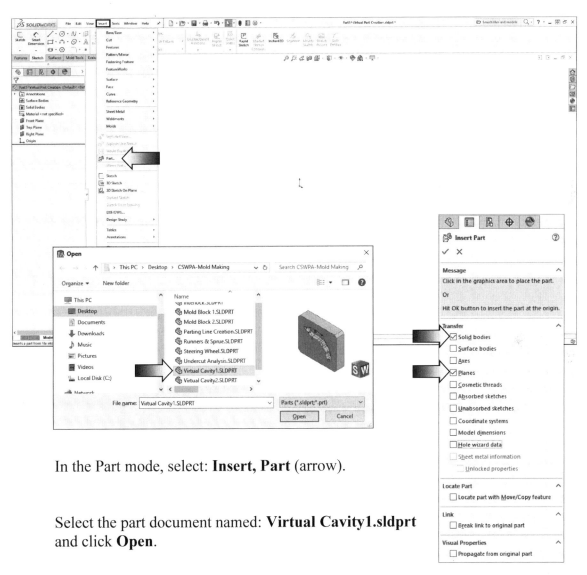

In the Part mode, select: **Insert, Part** (arrow).

Select the part document named: **Virtual Cavity1.sldprt** and click **Open**.

Enable the 2 checkboxes shown in the Insert Part dialog box and click **OK**. The part is placed on the Origin by default.

The part Virtual Cavity1 is inserted but not yet centered to match its assembly.

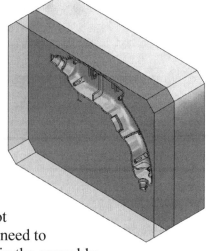

It will be moved to the exact location (X0, Y0, Z0) after the 2nd part is inserted.

6. Inserting the 2nd part:

Select **Insert, Part** once again.

Select the part document named: **Virtual Cavity2** (arrow) and click **Open**.

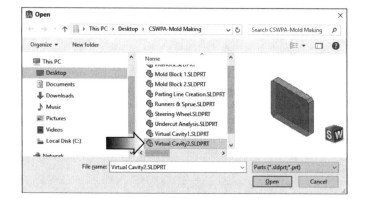

Enable the 2 options:

    * Solid Bodies

    * Planes

Click **OK** to place the part on the **Origin**.

When modeling components in the context of an assembly, or Top Down method, the origins of the components are often "off center" when they are opened in their own windows. In this case, the origins of the 2 solid bodies are not exactly the same as the one in the assembly. They need to be moved to match the position of the mold bases in the assembly.

7. Moving the solid bodies:

Select: **Insert, Features, Move/Copy** (arrow).

For Bodies to Move/Copy, select the 2 bodies **Virtual Cavity1** and **Virtual Cavity2** either from the graphics area or from the Feature tree.

In the Translate section, enter the following:

Delta X: **-1.141in**

Delta Y: **.280in**

Click **OK**.

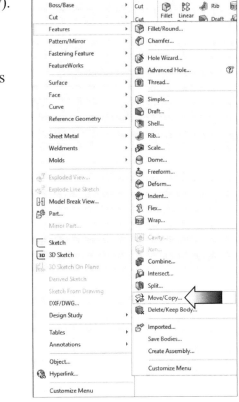

The 2 solid bodies are moved to the correct position (X0, Y0, Z0).

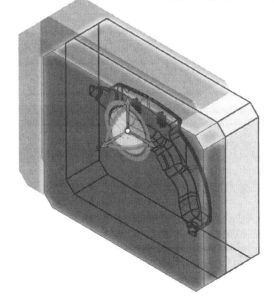

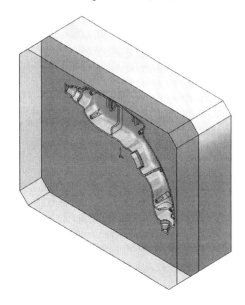

8. Creating an Intersect feature:

There are several ways to create the plastic part from the Core and Cavity block, we will try the **Intersect** option first.

Click the **Intersect** command 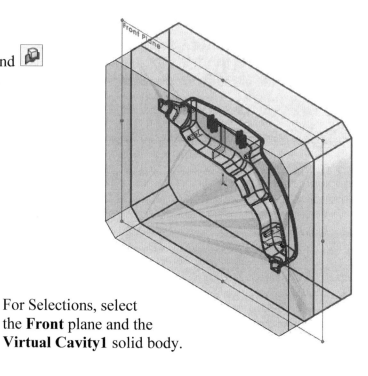 from the **Features** tool tab.

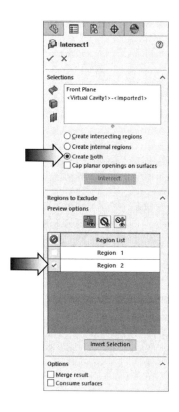

For Selections, select the **Front** plane and the **Virtual Cavity1** solid body.

Click the option: **Create Both** (arrow).

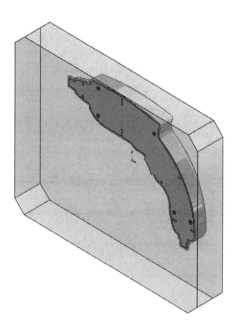

For Regions to Exclude, select: **Region 2** (arrow).

Clear the **Merge Result** and **Consume Faces** checkboxes.

Click **OK**.

9. Creating a Combine Subtract feature:

Next, we will use the **Combine Subtract** option to remove the block.

Select **Insert, Features, Combine**.

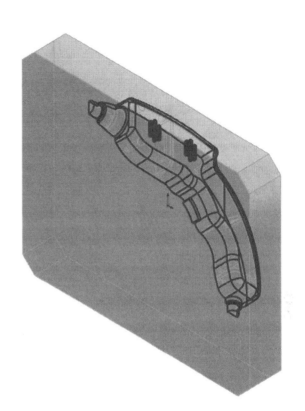

For Operation Type, select:
Subtract (arrow).

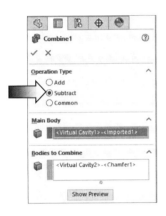

For Main Body, select the **Virtual Cavity1** solid body.

For Bodies to Combine, select the **Virtual-Cavity2** solid body.

Click **Show Preview** to see the result of the subtraction.

Click **OK**.

The plastic part is created from the Core and Cavity blocks.

10. Working in the Assembly level:

Select **Window, Virtual Part-Creation.sldasm** (arrow) to go back to the assembly document.

The hotkey **Control+Tab** also switches back and forth between the part and the assembly documents.

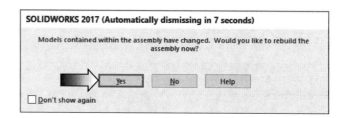

Click **Yes** in the message box to rebuild the assembly document.

The plastic part and the 2 mold blocks are centered about the assembly's origin.

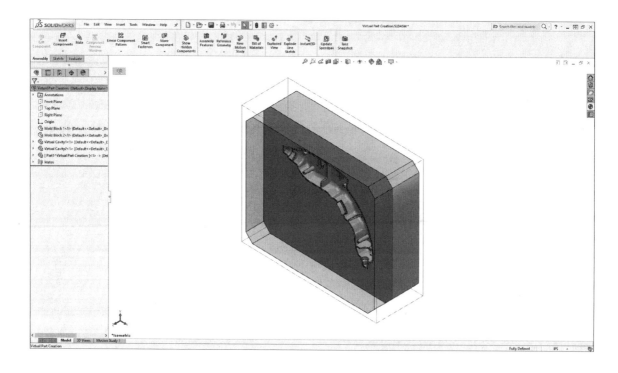

11. Assigning material:

Expand the Component named **Part1** to see the material option.

Right-click the Material option and select **Edit Material**.

Select **ABS** from the Plastic folder and click **OK**.

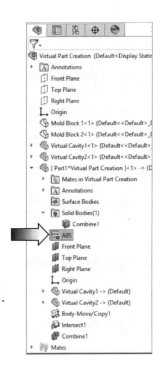

12. Calculating the mass of the part:

Switch to the **Evaluate** tab and click **Mass properties** ⚖.

Select **Part1** from the Feature tree to calculate its mass.

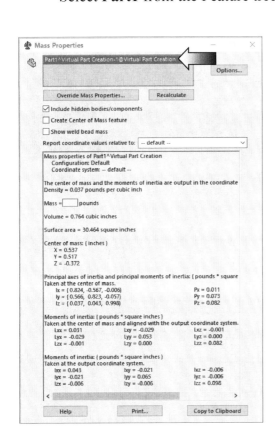

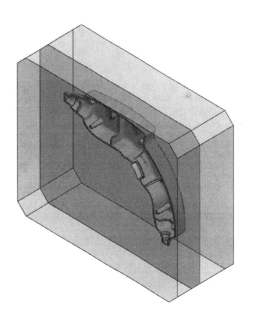

Enter the overall mass of the **Part1** here: _____ lbs.

13. Saving your work:

Select **File, Save**.

Save the document using the same name and allow to overwrite when prompted.

Close all documents.

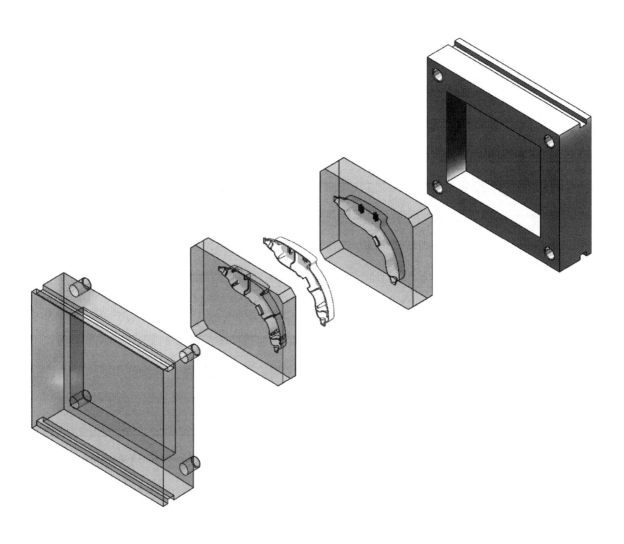

CHAPTER 3

CSWP — Advanced Weldments

CSWP – Advanced Weldments

The completion of the Certified SOLIDWORKS Professional Advanced Weldments (CSWPA-WD) exam proves that you have successfully demonstrated your ability to use the SOLIDWORKS tools for Weldments.

Employers can be confident that you understand the SOLIDWORKS tools that will aid in the design of Weldment components.
Recommended Training Courses: Weldments

Note: *You must use at least SOLIDWORKS 2010 for this exam. Any use of a previous version will result in the inability to open some of the testing files.*

Exam Length: 2 hours
Minimum Passing grade: 75%

Re-test Policy: There is a minimum 14 days waiting period between every attempt of the CSWPA-WD exam. Also, a CSWPA-WD exam credit must be purchased for each exam attempt.

All candidates receive electronic certificates and a personal listing on the CSWP directory when they pass.

Exam features hands-on challenges in many of these areas of SOLIDWORKS Weldment functionality:

Weldment Profile creation, Placing Weldment Profile in the Weldment Library, Basic and Advanced Weldment Part Creation and modification, Adding End Caps and Gussets, Using Trim & Extend command, 3D Sketch creation, Cut List Management and Cut List Creation in Weldment Drawings.

CSWP – Advanced Weldments

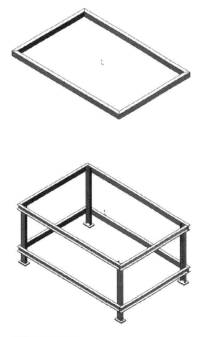

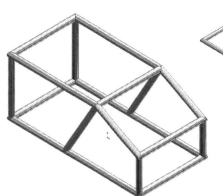

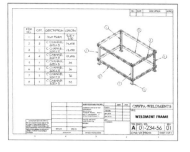

View Orientation Hot Keys:

Ctrl + 1 = Front View
Ctrl + 2 = Back View
Ctrl + 3 = Left View
Ctrl + 4 = Right View
Ctrl + 5 = Top View
Ctrl + 6 = Bottom View
Ctrl + 7 = Isometric View
Ctrl + 8 = Normal To Selection

Dimensioning Standards: **ANSI**

Units: **INCHES** – 3 Decimals

Tools Needed:

Weldment	Rectangle	Circle
End Cap	Structural Member	Trim/Extend
Cut List	Drawing	Mass Properties

CHALLENGE 1

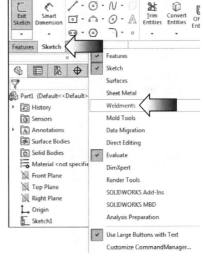

This challenge focuses on creating the weldment profiles and saving them in the Weldment Profiles Library.

1. Starting a new part document:

Click **File / New**.

Select the default **Part Template** and set the Units to **Inches**, **3 decimal places**.

Weldments functionality enables you to design a weldment structure as a single multibody part. You use 2D and 3D sketches to define the basic framework, and then create structural members containing groups of sketch segments. You can also add items such as gussets and end caps using tools on the Weldments toolbar.

2. Enabling the Weldments tab:

Right-click on the **Sketch** tab and select the **Weldments** tab (arrow).

When you create the first structural member in a part, a **Weldment** feature and a **Cut List** are created (arrow) and added to the FeatureManager design tree.

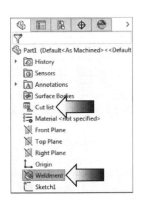

SOLIDWORKS also creates two default configurations in the ConfigurationManager: a parent configuration Default (As Machined) and a derived Configuration Default (As Welded).

The next step is to create the first weldment profile and save it in the Weldment profiles Library.

3. Sketching the 1ˢᵗ weldment profile:

Select the Front plane and open a **new sketch**.

Sketch **2 Rectangles** and add the dimensions as indicated to fully define the sketch.

Add the **Symmetric** relations where needed to reduce some of the redundant dimensions.

Add the **Sketch Point** as indicated to help locate the profile later on.

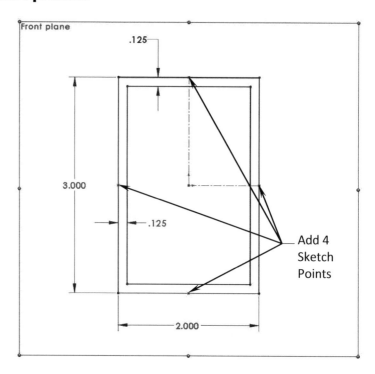

4. Saving the profile:

<u>Exit</u> the Sketch. <u>Select</u> **Sketch1** and **save** it under the following directories:

C: Program Files/Solid-Works Corp/ SolidWorks/ Lang/English/ Weldment Profiles/Ansi Inch/ Rectangular Tube.

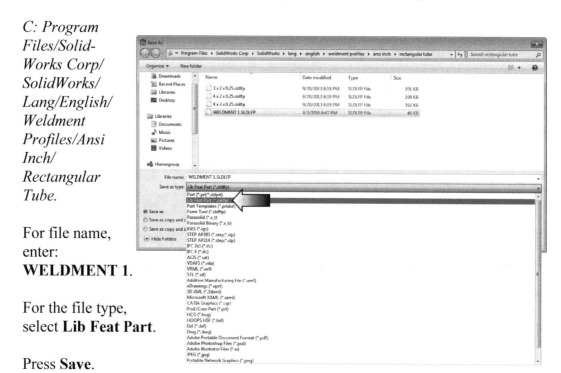

For file name, enter: **WELDMENT 1.**

For the file type, select **Lib Feat Part.**

Press **Save.**

5. Creating the 2ⁿᵈ weldment profile:

Click **File / New / Part**.

Select the default **Part Template** and set the Units to **Inches, 3 decimal places**.

6. Sketching the profile:

Open a **new sketch** on the Front plane.

Sketch **2 circles** and add the diameter dimensions to fully define the sketch.

Add **4 Sketch Points** as noted to help locate the profile later on.

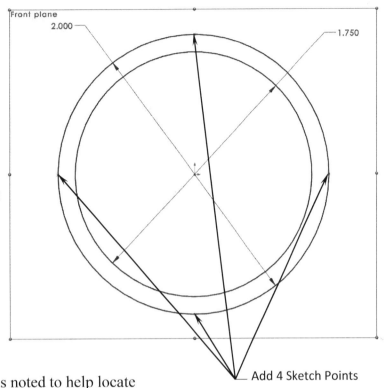

Add 4 Sketch Points

7. Saving the profile:

<u>**Exit**</u> the sketch. <u>Select</u> **Sketch1** and <u>**save**</u> it under the same directory as the last one.

For the file name, enter: **WELDMENT 2**.

For the file type, select **Lib Feat Part**.

Press **Save** and <u>close</u> all documents.

8. Showing the Weldment Profiles Folder:

The option "Show Folder For" lists the file types for which you can define a search path. You specify the folders to search for different types of documents. Folders are searched in the order in which they are listed.

The Weldment Profiles are saved in the following directories: **C:\Program Files\ SOLIDWORKS Corp\SOLIDWORKS\Lang\English\Weldment Profiles.**

Select **Options / File Locations / Show Folders For / Weldment Profiles** (arrows) and click **OK**.

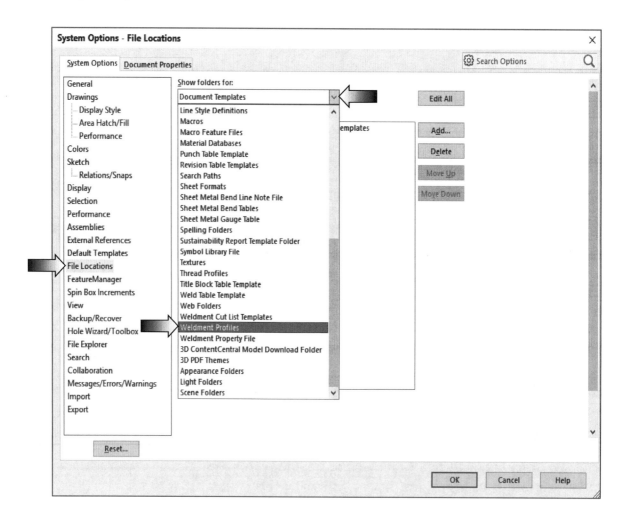

*Note: If the Weldment Profiles folder is not visible, click **Add** and browse to the Weldment Profiles directory listed above.*

9. Creating a new weldment part:

Click **File / New / Part**.

Select the default **Part Template** and set the Units to **Inches, 3 decimal places**.

Select the <u>Top</u> plane and open a **new sketch**.

Sketch a **Center Rectangle** as shown and add the height and width dimensions.

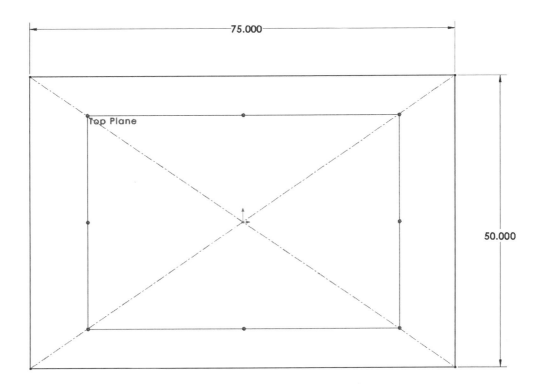

Exit the sketch.

Click the **Weldment** command (arrow) to add the Weldment and the Cut List features to the FeatureManager tree.

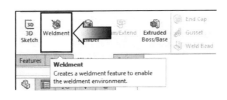

Two default configurations are added in the Configuration-Manager: a parent configuration Default [As Machined] and a derived configuration Default [As Welded].

10. Adding Structural Members:

The weldment profiles that were created earlier will be used to create the weldment structural members.

Click the **Structural Member** command (arrow).

For Standard, select **Ansi Inch**.

For Type, select **Rectangular Tube**.

For Size, select **WELDMENT1**.

For Group1, select the **4 lines** in the graphics area.

For Apply Corner Treatment, use the default **End Miter** option.

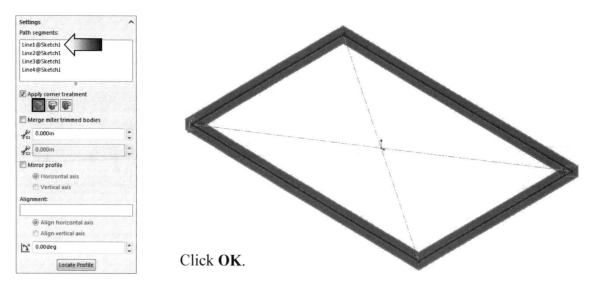

Click **OK**.

11. Applying material:

You can apply a material to a part, or one or more bodies of a multibody part. In the FeatureManager design tree, expand the Solid Bodies folder, right-click a body and select Material. To affect several bodies, Control+select them before right-clicking.

Right-click the **Material** option on the Feature Manager tree (arrow) and select: **Plain Carbon Steel**.

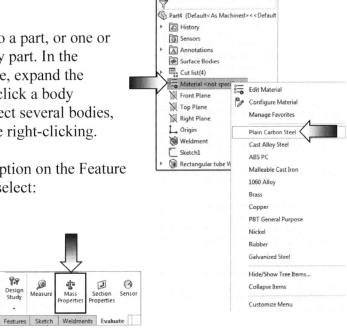

12. Calculating the mass:

Click **Mass Properties** on the **Evaluate** tab.

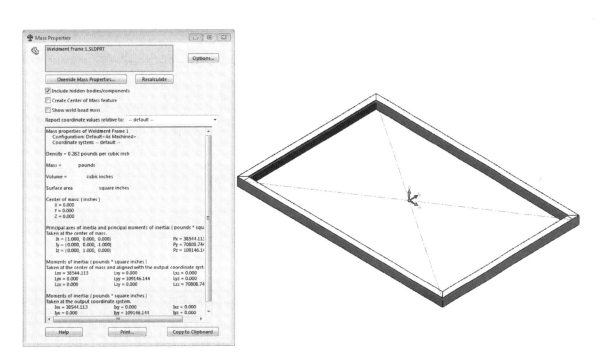

Enter the total mass of the part here: _____ lbs.

Save the part file as **Weldment Frame 1** but keep the part file open.

13. Changing the Structural Members:

Delete the Rectangular Tube feature from the FeatureManager tree but keep the **Sketch1**. We will apply the 2nd weldment profile to it.

Click the **Structural Member**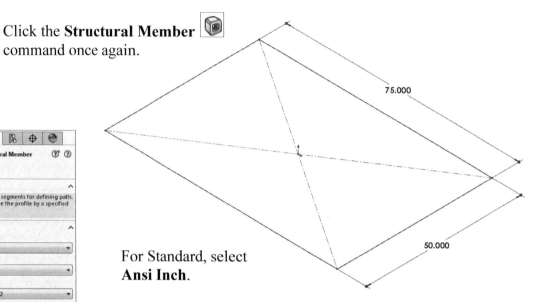
command once again.

For Standard, select **Ansi Inch**.

For Type, select **Pipe**.

For Size, select **WELDMENT 2**.

For Group1, select the **4 Lines** in the graphics area.

For Apply Corner Treatment, use the default **End Miter** option.

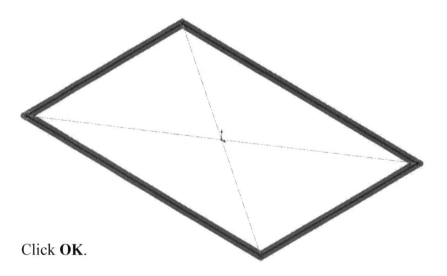

Click **OK**.

14. Calculating the mass:

Use the same material as the last weldment part:
Plain Carbon Steel.

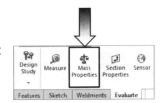

Switch to the **Evaluate** tab and click:
Mass Properties.

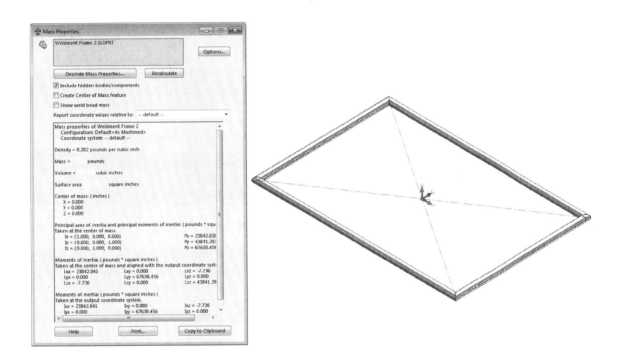

Enter the mass here: _____ lbs.

Save the part file as **Weldment Frame 2**.

Close all documents.

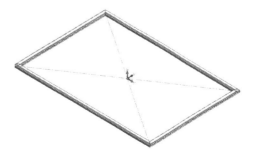

CHALLENGE 2

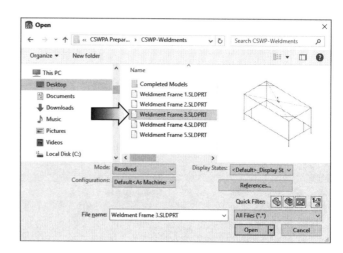

1. Opening a part document:

Select **File / Open**.

Browse to the Training Folder and open a part document named:
Weldment Frame 3.sldprt.

This challenge focuses on creating a weldment part by applying the correct weldment profiles, trimming, and treating the corners.

2. Adding Structural Members:

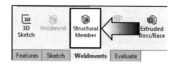

Click the **Weldments** command to add it to the tree.

Click **Structural Member** (arrow) and set the following:
> Standard: **Ansi Inch**.
> Type: **Rectangular Tube**.
> Size: **3 X 2 X 0.25**.

Select the **4 lines** from the graphics area. for Group1.

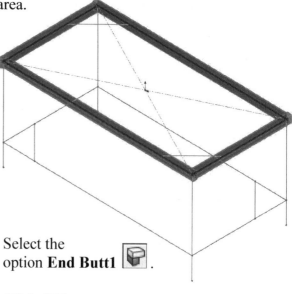

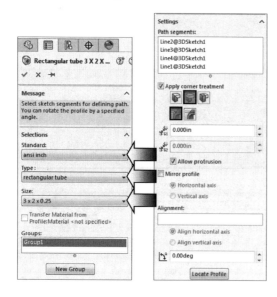

Select the option **End Butt1** .

Click **OK**.

A group is a collection of related segments in a structural member. You configure a group to affect all its segments without affecting other segments or groups in the structural member.

There are 2 types of groups:

Contiguous Group: A continuous contour of segments joined end-to-end. You can control how the segments join to each other. The end point of the group can optionally connect to its beginning point.

Parallel Group: A discontinuous collection of parallel segments. Segments in the group cannot touch each other.

3. Adding Structural Members to Group 2:

Click the **Structural Member** command again.

Use the <u>same selections</u> as Group 1 and select the **4 lines** below Group 1.

Keep the Corner Treatment at its default settings, including the **End Butt 1** .

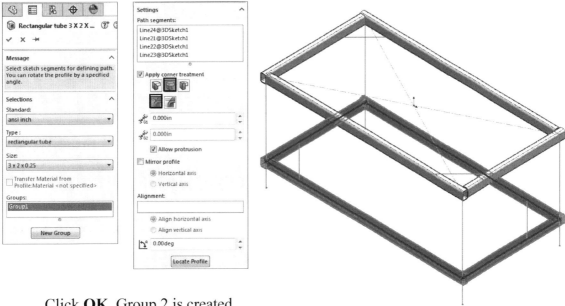

Click **OK**. Group 2 is created.

Note: If all groups are created within the same operation, SOLIDWORKS can automatically trim the intersections, but we will be using the manual trim for this lesson.

4. Creating the Parallel Group:

Click the **Structural Member** command .

Use the <u>same selections</u> as the last 2 groups (Ansi Inch, Rectangular 3x2x0.25).

Select the **4 Vertical Lines** as shown.

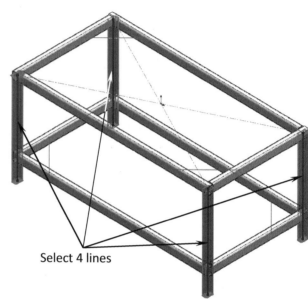

Select 4 lines

Click **OK**.

A Parallel Group, Group 3, is created.

The weldment part has some interference at this point (circled).

The tubes will be trimmed to the correct lengths in the next couple of steps.

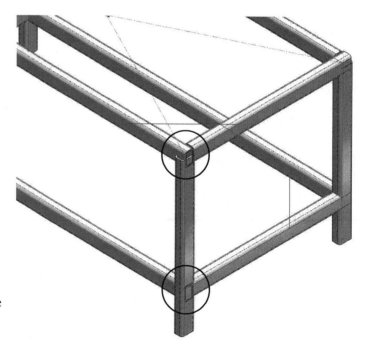

5. Trimming the Parallel Group:

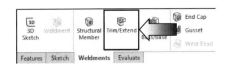

Click the **Trim/Extend** command .

For Corner Type, use the default **End Trim** .

For Bodies to be Trimmed, select the **4 Vertical Tubes**.

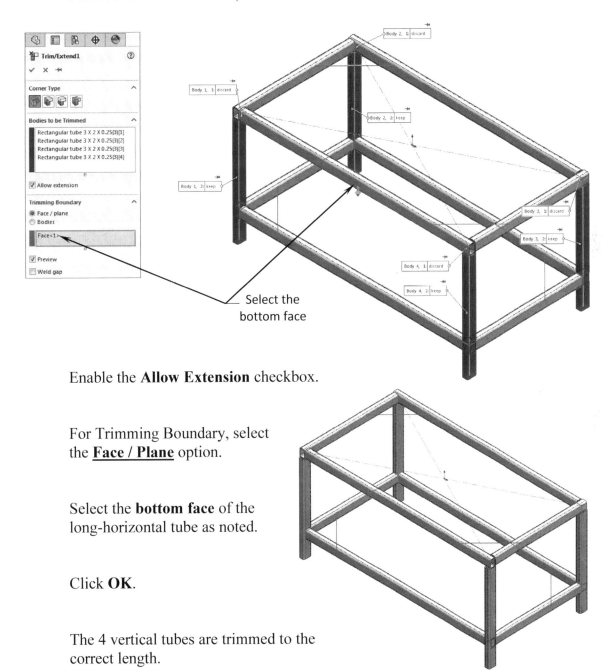

Select the bottom face

Enable the **Allow Extension** checkbox.

For Trimming Boundary, select the **Face / Plane** option.

Select the **bottom face** of the long-horizontal tube as noted.

Click **OK**.

The 4 vertical tubes are trimmed to the correct length.

6. Trimming the Group 2:

Click the **Trim/Extend** command .

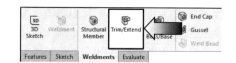

For Corner Type, use the default **End Trim** .

For Bodies to be Trimmed, select the **4 horizontal tubes** in Group 2.

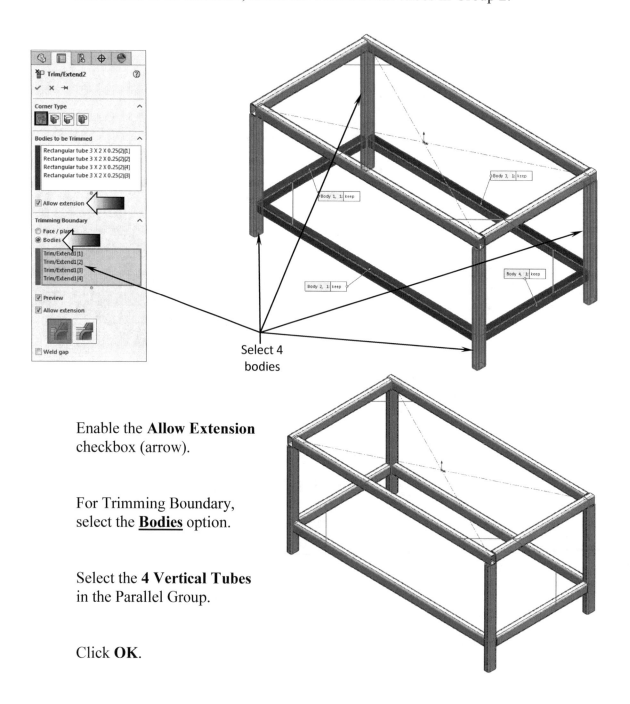

Select 4 bodies

Enable the **Allow Extension** checkbox (arrow).

For Trimming Boundary, select the **Bodies** option.

Select the **4 Vertical Tubes** in the Parallel Group.

Click **OK**.

7. Applying material:

Right-click the **Material** option on the Feature Manager tree and select:
Plain Carbon Steel.

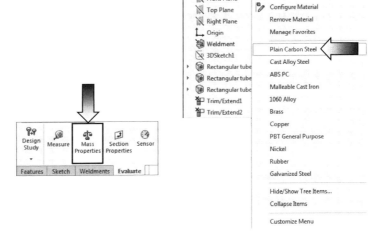

8. Calculating the mass:

Click **Mass Properties** on the **Evaluate** tab.

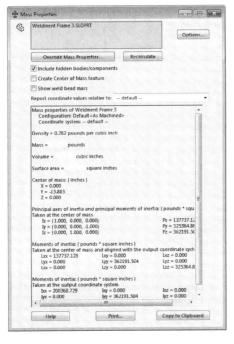

Enter the mass of the part here: _____ lbs.

9. Creating the Upper Braces:

Click the **Structural Member** command .

Use the <u>same selections</u> as the other 3 groups (Ansi Inch, Rectangular 3x2x0.25).

Select the **2 angled lines** on the <u>top</u> of the main sketch.

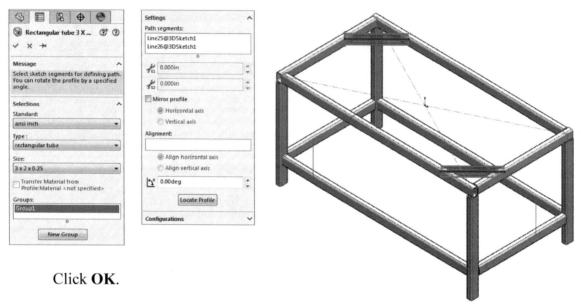

Click **OK**.

10. Trimming the 1st Upper Brace:

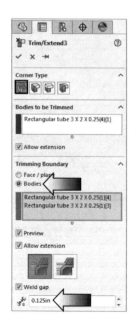

Trim the Upper Brace using the **End Trim** and **Bodies** options.

<u>Enable</u> the **Weld Gap** checkbox (arrow) and enter **.125in** for gap.

Click **OK**.

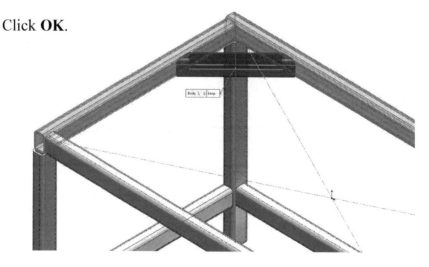

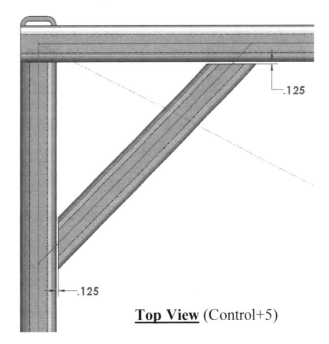

The first Upper Brace is shown from the Top view with **.125in** gaps on either ends.

.125

.125

Top View (Control+5)

11. Trimming the 2ⁿᵈ Upper Braces:

Trim the Upper Brace using the **End Trim** and **Bodies** options.

Enable the **Weld Gap** checkbox and enter: **.125in** for gap.

Enable **Allow Extension**.

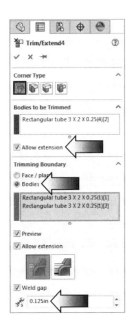

Click **OK**.

Top View (Control+5)

.125

The 2nd Upper Brace is shown from the Top view with **.125in** gaps on either end.

.125

12. Updating the Cut list:

The Cut List should be updated after a new group is added to the weldment part. Identical items are grouped together in Cut-List-Item subfolders.

The icon indicates the Cut list is up to date.

The icon indicates the Cut list should be updated.

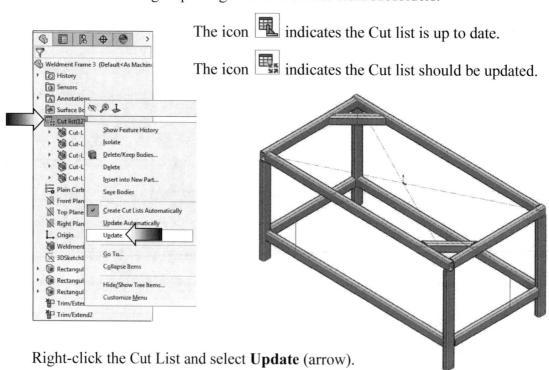

Right-click the Cut List and select **Update** (arrow).

13. Creating the Lower Braces:

Click the **Structural Member** command .

Use the <u>same selections</u> as the other 4 groups (Ansi Inch, Rectangular 3x2x0.25).

Select the **2 angled lines** on the <u>bottom</u> of the main sketch.

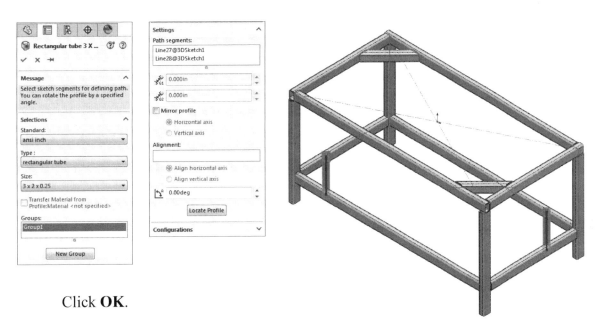

Click **OK**.

14. Trimming the 1st Lower Brace:

Trim the Lower Brace using the **End Trim** and **Bodies** options.

<u>Enable</u> the **Weld Gap** checkbox (arrow).

<u>Enable</u> the **Allow Extension** checkbox.

Enter: **.125in** for gap.

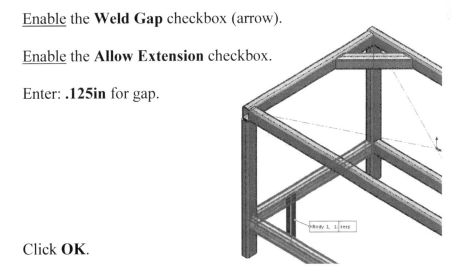

Click **OK**.

.125

The first Lower Brace is shown from the Bottom view with **.125in** gaps on either end.

.125

Bottom View (Control+6)

15. Trimming the 2nd Lower Brace:

Trim the Lower Brace using the **End Trim** and **Bodies** options.

<u>Enable</u> the **Weld Gap** checkbox and enter **.125in** for gap.

<u>Enable</u> **Allow Extension**.

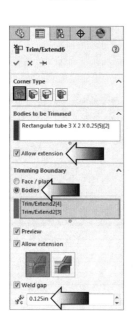

Click **OK**.

The second Lower Brace is shown from the Top view with **.125in** gaps on either end.

Bottom View (Control+6)

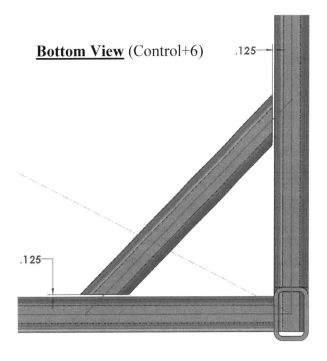

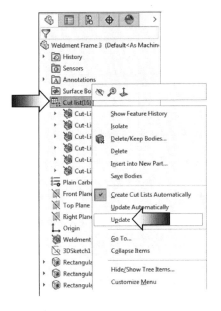

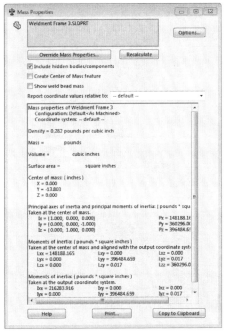

Update the Cut List once again to group the identical structural members into sub-folders.

Enter the mass of the entire weldment part here: _____ lbs.

16. Adding End Caps:

End caps can be added, including internal end caps to close-off open structural members.

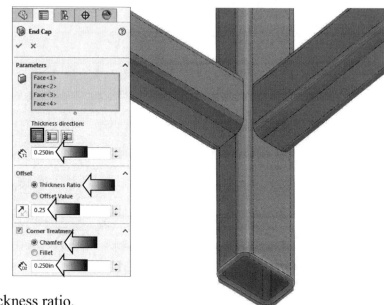

To apply fillets or chamfers to all end caps, under Corner Treatment, select Fillet or Chamfer.

You can offset an end cap from the inside face of a structural member by specifying a decimal Offset value in addition to the Thickness ratio.

Click the **End Cap** command 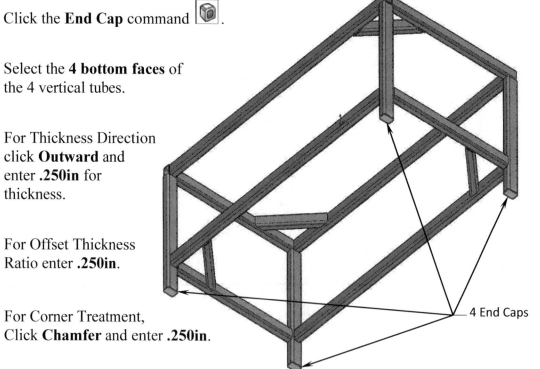.

Select the **4 bottom faces** of the 4 vertical tubes.

For Thickness Direction click **Outward** and enter **.250in** for thickness.

For Offset Thickness Ratio enter **.250in**.

For Corner Treatment, Click **Chamfer** and enter **.250in**.

Click **OK**.

4 End Caps

17. Calculating the final mass:

Select the **Mass Properties** command from the **Evaluate** tab.

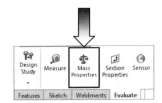

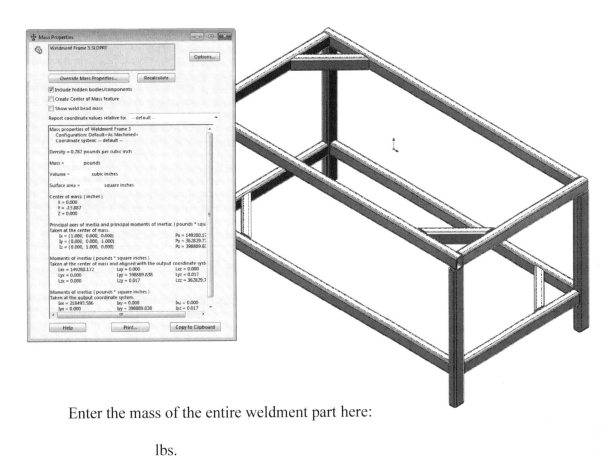

Enter the mass of the entire weldment part here:

_____ lbs.

18. Saving your work:

Select **File / Save As**.

Enter **Weldment Frame 3** for the name of the file.

Click **Save** and overwrite the old file with your completed one.

Close all documents.

CHALLENGE 3

1. Opening a part document:

Select **File / Open**.

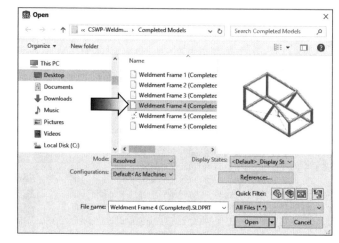

Browse to the Training Folder and open a part document named :
Weldment Frame 4.sldprt.

This challenge focuses on creating a weldment part from a 3D Sketch and trimming the Structural Members.

2. Creating a 3D Sketch:

Click the **Weldment** command to add a Weldment feature to the tree.

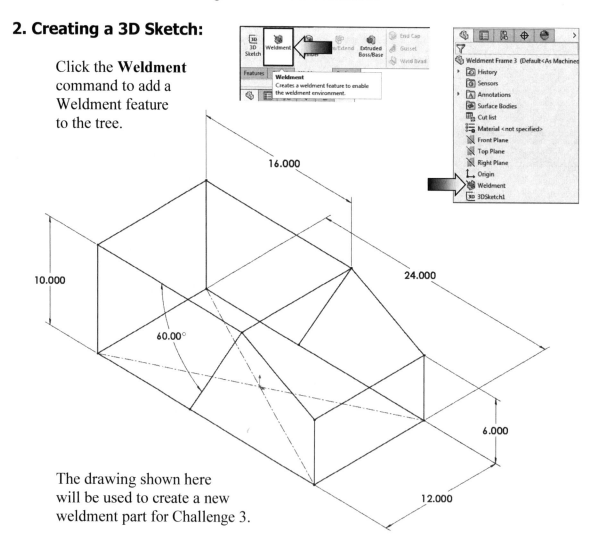

The drawing shown here will be used to create a new weldment part for Challenge 3.

Select **3D Sketch** (arrow) below the Sketch drop-down.

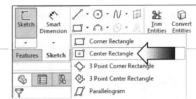

Select the **Center Rectangle** command, hold the **Control** key and click the **Top** plane (arrow).

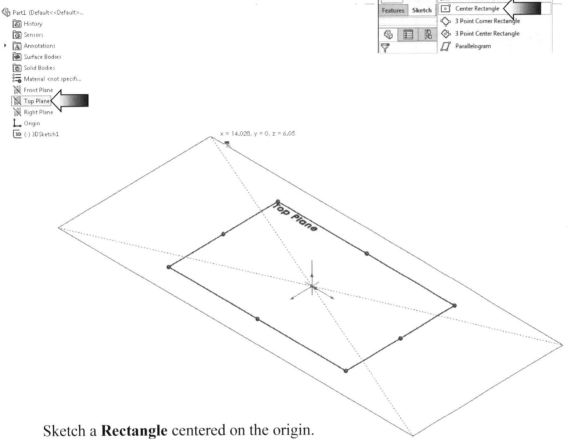

Sketch a **Rectangle** centered on the origin.

*An alternative way to sketch the same rectangle on the Top plane is to click the Origin to start the rectangle and push the **TAB** key once or twice while looking out for the **ZX** indicator that appears next to the mouse cursor, then move the cursor outward and click to complete the rectangle.*

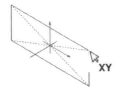

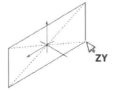

Add the **Along Z** (Vertical) relation to the line on the left side of the rectangle.

Add the **Along X** (Horizontal) to the line at the bottom.

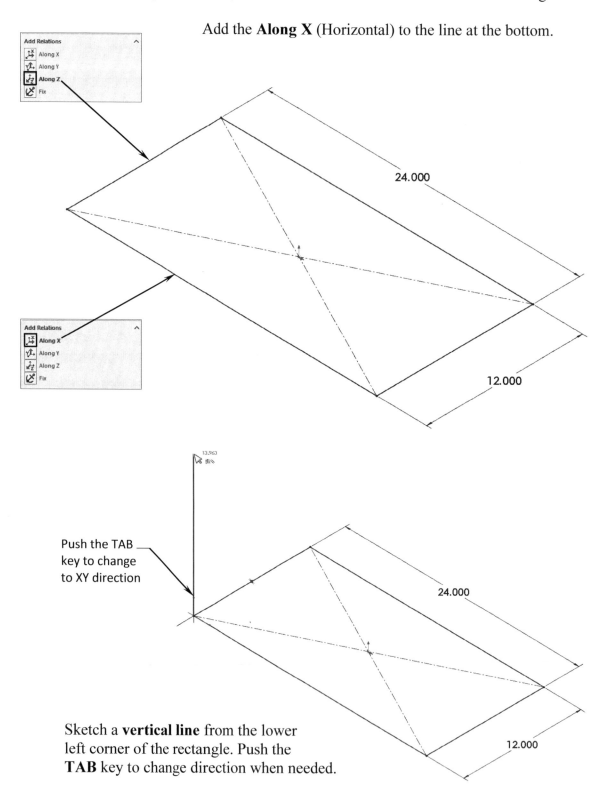

Push the TAB
key to change
to XY direction

Sketch a **vertical line** from the lower
left corner of the rectangle. Push the
TAB key to change direction when needed.

Sketch 3 more vertical **Lines** and add an **Equal relation** to <u>each pair</u> as noted.

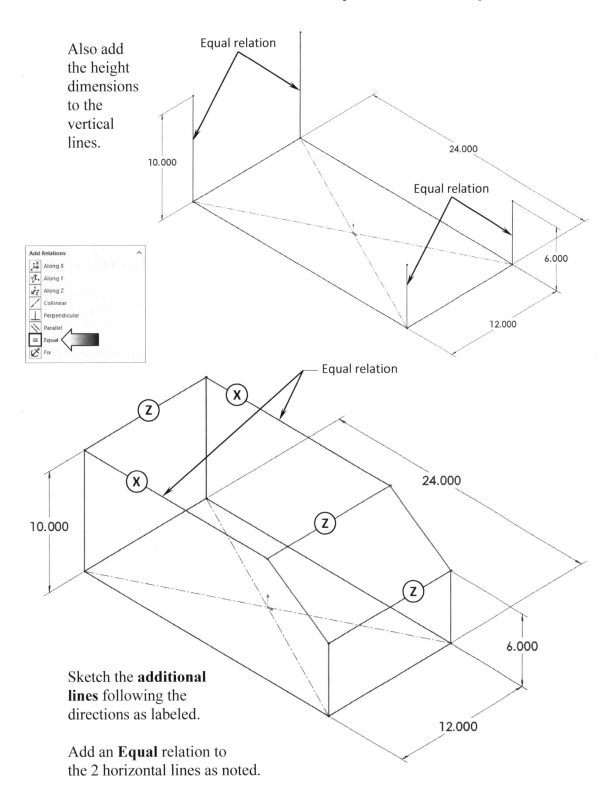

Also add the height dimensions to the vertical lines.

Equal relation

10.000

24.000

Equal relation

6.000

12.000

Add Relations

Along X
Along Y
Along Z
Collinear
Perpendicular
Parallel
= Equal
Fix

Equal relation

Z X

X

10.000

24.000

Z

Z

Z

6.000

Sketch the **additional lines** following the directions as labeled.

12.000

Add an **Equal** relation to the 2 horizontal lines as noted.

Sketch the **two 60 degree Lines** and add a **Parallel** relation to constrain them.

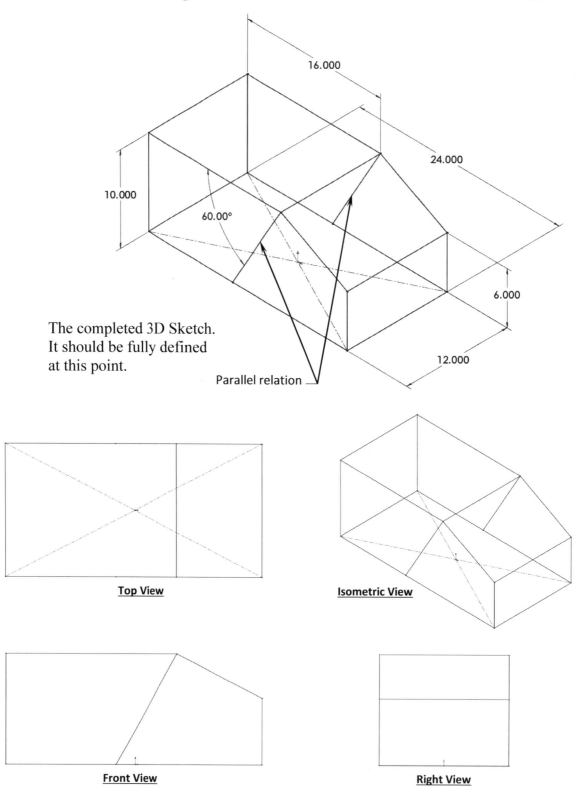

The completed 3D Sketch.
It should be fully defined
at this point.

Parallel relation

Top View

Isometric View

Front View

Right View

Exit the 3D Sketch (or press Control+Q).

3. Adding Structural Members:

Click the **Structural Member** command .

Select the following: * **Ansi Inch**   * **Pipe**   * **0.5 sch 40**

Select the **6 Lines** on top of the 3D Sketch.

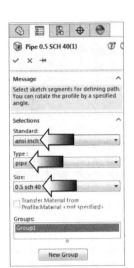

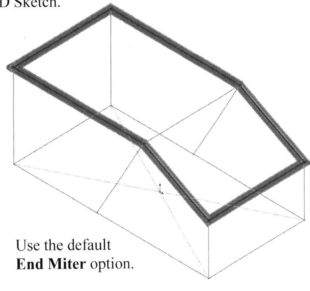

Use the default
End Miter option.

Click **OK**.

<u>Repeat</u> step 3 and add the same Structural Members to the **4 lines** at the bottom of the 3D Sketch.

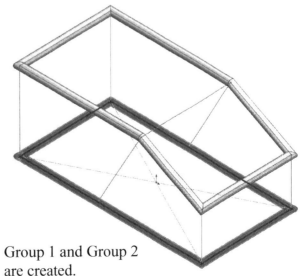

Group 1 and Group 2
are created.

4. Creating the Parallel Group:

Click the **Structural Member** command .

Use the same selections as the last 2 groups (Ansi Inch, Pipe, 0.5 sch 40) and select the **four vertical Lines** from the graphics area.

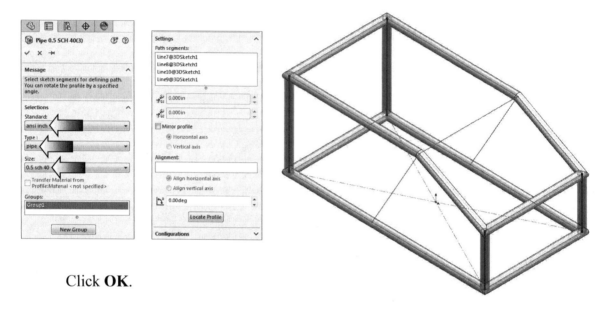

Click **OK**.

There are some interferences between the structural members. They will be corrected and trimmed to the lengths in the next couple of steps.

Repeat the step above and add the same structural member to the **vertical line** in the middle of the 3D Sketch.

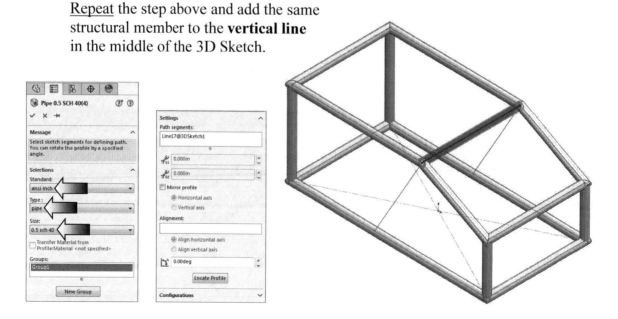

5. Creating the last group:

Click the **Structural Member** command .

Use the <u>same selections</u> as the last 4 groups (Ansi Inch, Pipe, 0.5 sch 40) and select the **two 60 degrees lines** from the graphics area.

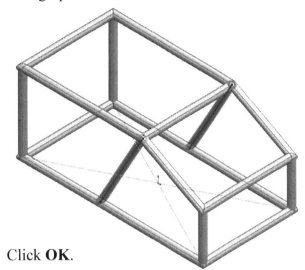

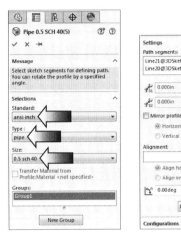

Click **OK**.

6. Trimming the 1st tubes:

Click **Trim/Extend** .

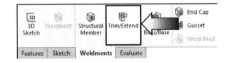

For Bodies to be Trimmed, select the **vertical tube** as noted and <u>clear</u> the **Allow Extension** checkbox.

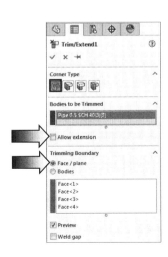

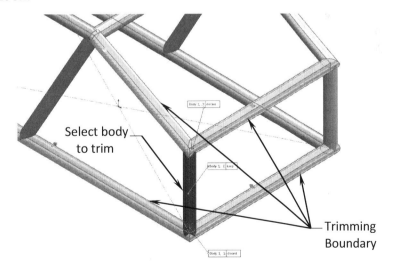

Select body to trim

Trimming Boundary

For Trimming Boundary, select the **Face/Plane** option and select the **4 structural members** as noted.

7. Trimming the 2nd tubes:

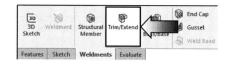

Click **Trim/Extend** .

For Bodies to be Trimmed, select the **vertical tube** on the far right as noted. <u>Uncheck</u> the **Allow Extension** checkbox.

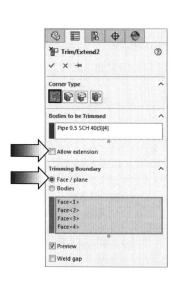

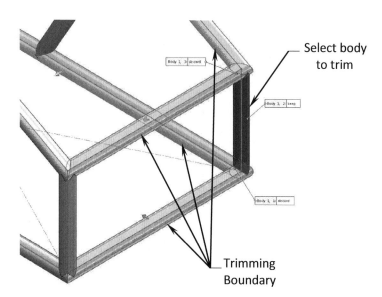

For Trimming Boundary, select the **Face/Plane** option and select the **4 structural members** as noted.

8. Trimming the 3rd tubes:

<u>Repeat</u> the step above and trim the vertical tube on the front-left as noted.

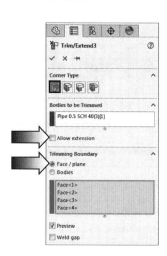

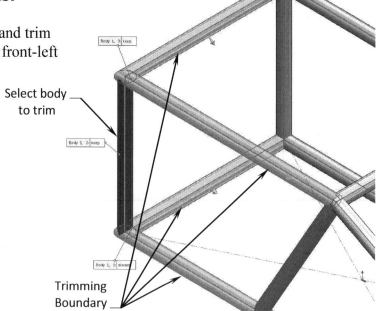

9. Trimming the 4th tubes:

Click **Trim/Extend** .

For Bodies to be Trimmed, select the **fourth vertical tube** on the back side.

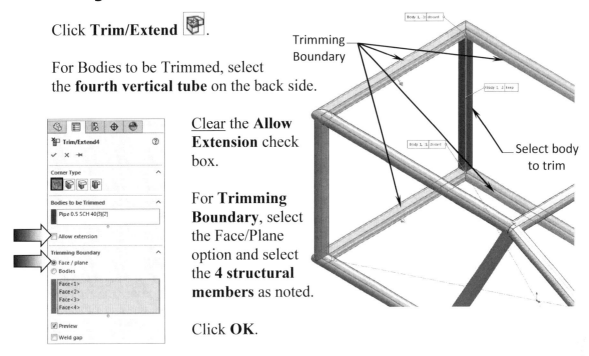

<u>Clear</u> the **Allow Extension** check box.

For **Trimming Boundary**, select the Face/Plane option and select the **4 structural members** as noted.

Click **OK**.

10. Trimming the 5th tubes:

Click **Trim/Extend** .

For Bodies to be Trimmed, select the two **60° tubes**.

For Trimming Boundary, select the **6 structural members** as indicated.

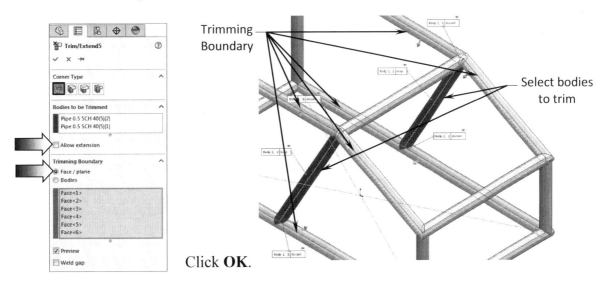

Click **OK**.

11. Trimming the 6th tubes:

Click **Trim/Extend** 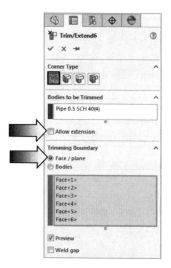.

For Bodies to be Trimmed, select the **support tube** in the middle of the weldment.

For Trimming Boundary, select the **6 structural members** as indicated.

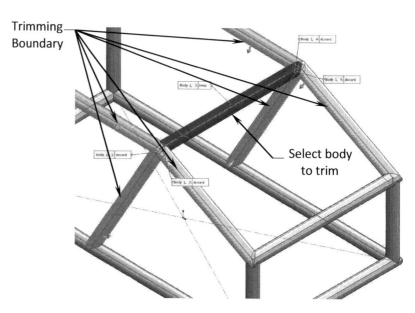

Trimming Boundary

Select body to trim

Click **OK**.

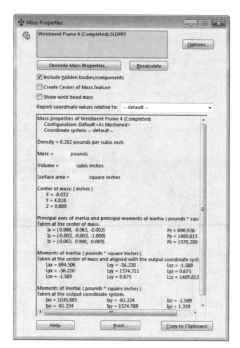

Assign the material **Plain Carbon Steel** to the entire weldment part.

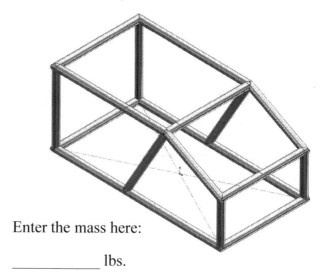

Enter the mass here:

_____ lbs.

CHALLENGE 4

1. Opening a part document:

Select **File / Open**.

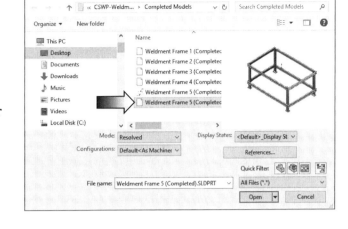

Browse to the Training Folder
and open the part document
named :
Weldment Frame 5.sldprt.

This challenge focuses on
Cut List Management, and
Cut List Creation in a Weldment Drawing.

2. Assigning material:

Right-click the **Material** option and select
Plain Carbon Steel.

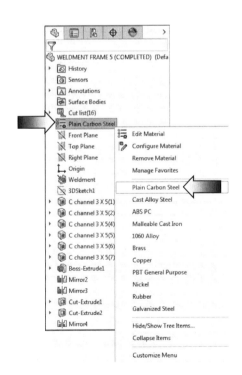

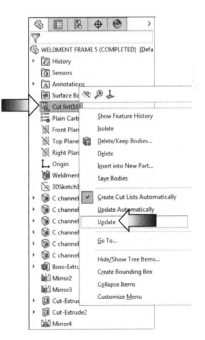

Right-click the Cut List and select **Update**. The
identical members will be grouped into subfolders.

3. Renaming a folder:

The Foot Plates were added to the weldment part as extruded features. They should be renamed to make it more descriptive on the Cut List.

Expand the Cut List and <u>Rename</u> the folder to **Foot Plates** (arrow).

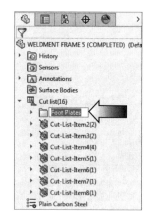
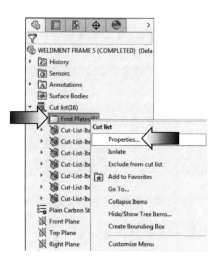

4. Modifying the Properties:

Right-click the <u>Foot Plate folder</u> and select **Properties**.

Click in line item 3 and enter **DESCRIPTION** (arrow). Enter **FOOT PLATES** under Value / Text Expression.

In line item 4, enter **LENGTH** (arrow). Enter: **5.00 X 3.00**.

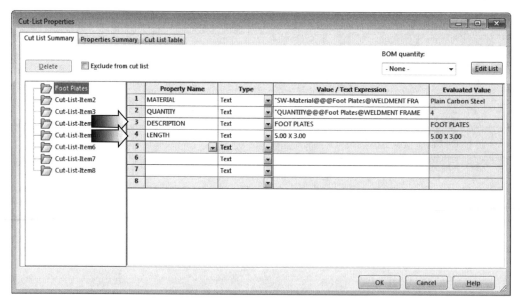

		Property Name	Type		Value / Text Expression	Evaluated Value
	1	MATERIAL	Text	▼	"SW-Material@@@Foot Plates@WELDMENT FRA	Plain Carbon Steel
	2	QUANTITY	Text	▼	"QUANTITY@@@Foot Plates@WELDMENT FRAME	4
	3	DESCRIPTION	Text	▼	FOOT PLATES	FOOT PLATES
	4	LENGTH	Text	▼	5.00 X 3.00	5.00 X 3.00
	5		Text	▼		
	6		Text	▼		
	7		Text	▼		
	8			▼		

Click **OK** and <u>Save</u> the model.

5. Transferring to a drawing:

Select **File / Make Drawing From Part** (arrow).

Select the default **Drawing Template** and click **OK**.

For this challenge, use the default **A-Size** paper.

Set the Projection to **3rd Angle**.

Set Drafting Standards to **ANSI**.

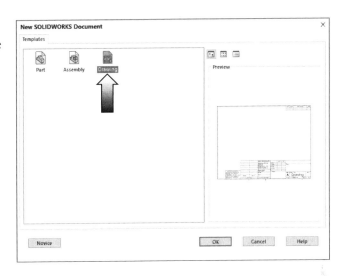

6. Adding a drawing view:

Expand the **View Palette** (arrow).

Drag and drop the **Isometric view** to the drawing approximately as shown.

Change the view **Scale** to **1:24**.

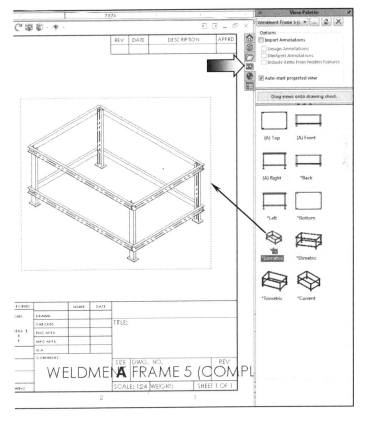

7. Inserting the Weldment Cut List:

Click the dotted border of the
Isometric view to activate it.

Switch to the **Annotation** tab.

Select the **Weldment Cut List** under
the Tables drop-down (arrow).

Keep all the default settings as they are and click **OK**.

Place the Weldment Cut List on the left side of the isometric view.

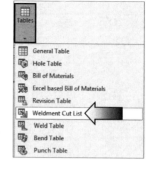

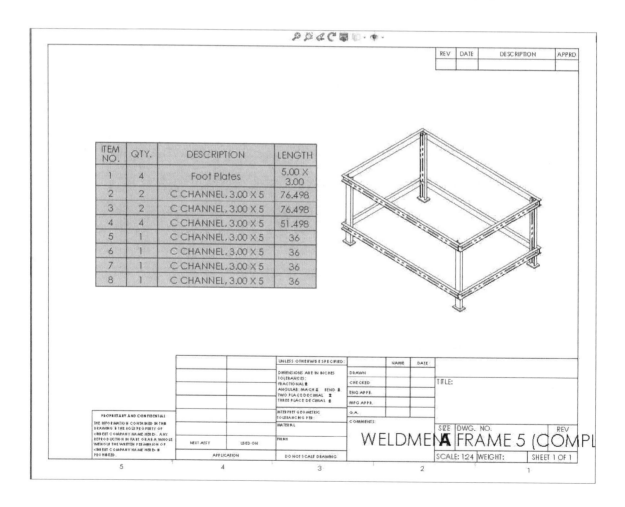

ITEM NO.	QTY.	DESCRIPTION	LENGTH
1	4	Foot Plates	5.00 X 3.00
2	2	C CHANNEL, 3.00 X 5	76.498
3	2	C CHANNEL, 3.00 X 5	76.498
4	4	C CHANNEL, 3.00 X 5	51.498
5	1	C CHANNEL, 3.00 X 5	36
6	1	C CHANNEL, 3.00 X 5	36
7	1	C CHANNEL, 3.00 X 5	36
8	1	C CHANNEL, 3.00 X 5	36

8. Adding Balloons:

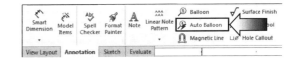

Click **Auto Balloons** .

Clear **Insert Magnetic Lines** and change the Balloon Settings to **1 Character**.

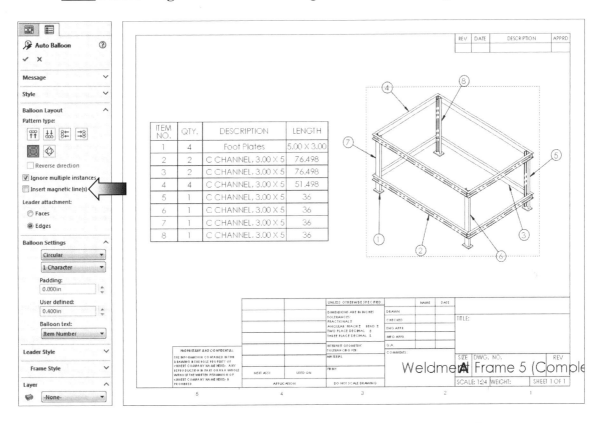

Click **OK** and re-arrange the balloons to look similar to the ones displayed above.

Locate **Item No. 4**, find the Length of the **C Channel, 3.00 X 5** and enter it here:

_____ in.

Save your work as **Weldment Frame 5.slddrw.**

Close all documents.

ITEM NO.	QTY.	DESCRIPTION	LENGTH
1	4	Foot Plates	5.00 X 3.00
2	2	C CHANNEL, 3.00 X 5	76.498
3	2	C CHANNEL, 3.00 X 5	76.498
4	4	C CHANNEL, 3.00 X 5	51.498
5	1	C CHANNEL, 3.00 X 5	36
6	1	C CHANNEL, 3.00 X 5	36
7	1	C CHANNEL, 3.00 X 5	36
8	1	C CHANNEL, 3.00 X 5	36

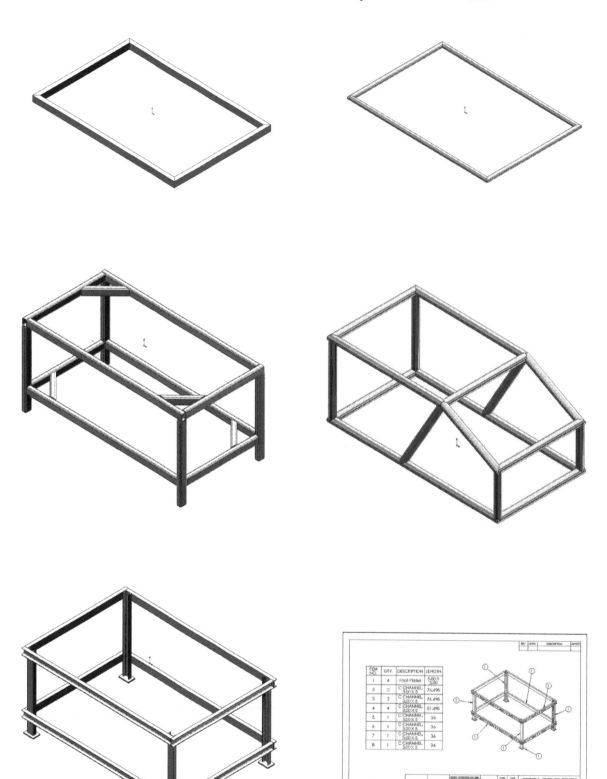

CHAPTER 4

CSWP — Advanced Sheet Metal

PROFESSIONAL
Sheet Metal

3S SOLIDWORKS

CSWP – Advanced Sheet Metal

The completion of the Certified SOLIDWORKS Professional
Advanced Sheet Metal (CSWPA-SM) exam shows that you have successfully
demonstrated your ability to use SOLIDWORKS Sheet Metal tools.

Employers can be confident that an individual with this certification understands
the set of SOLIDWORKS tools that aid in the design of sheet metal components.
Recommended training courses: SOLIDWORKS Essentials and Sheet Metal.

Note: A new version of the Sheet Metal exam in English has been released and
requires the use of SOLIDWORKS 2011 or later. For now, all non-English
versions of the test will still be the older Sheet Metal exam (2006 - 2010 version).

Exam Length: 2011 Version (English): 90 minutes
　　　　　　　2006 - 2010 Version (Non-English): 2 hours

Minimum Passing grade: 75%

Re-test Policy: There is a minimum 14 day waiting period between every attempt
of the CSWPA-SM exam. Also, a CSWPA-SM exam credit must be purchased for
each exam attempt. All candidates receive electronic certificates and a personal
listing on the CSWP directory when they pass.

Exam features hands-on challenges in many of these areas of SOLIDWORKS
Sheet Metal functionality such as:

Create Flanges, Closed Corners, Gauge Tables, Bend Calculation Options,
Hem, Jogs, Sketched Bends, Fold and Unfold, Forming Tools, Flatten,
Convert to Sheet Metal, and Sheet Metal Cut List Properties.

MS-Excel should be installed on the computer you are running SOLIDWORKS
on to ensure that your Gauge Table functionality works properly for the exam.

CSWP – Advanced Sheet Metal

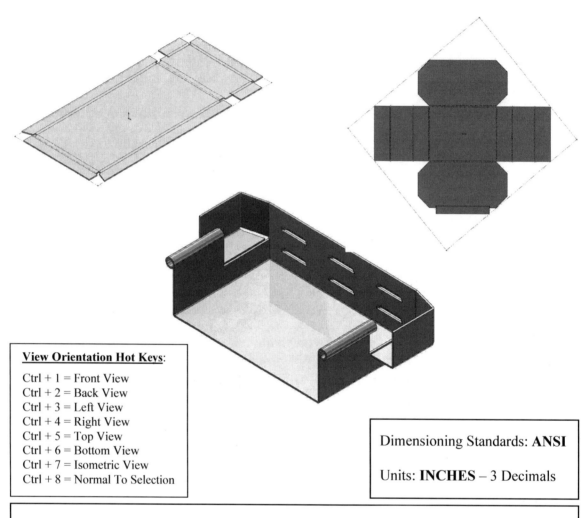

Dimensioning Standards: **ANSI**

Units: **INCHES** – 3 Decimals

Tools Needed:

 Convert to Sheet Metal

 Base Flange

 Hem

 Edge Flange

 Miter Flange

 Mirror

 Linear Pattern

 Forming Tool

 Cut List

CHALLENGE 1

1. Opening a part document:

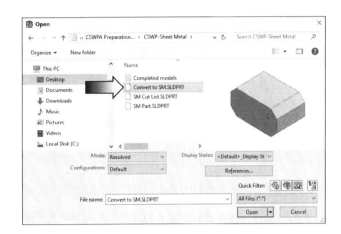

Select **File / Open**.

Browse to the Training Folder and open a part document named: **Convert to SM.sldprt**.

This challenge examines your skills on converting a solid part into a sheet metal model.

2. Enabling the Sheet Metal tab:

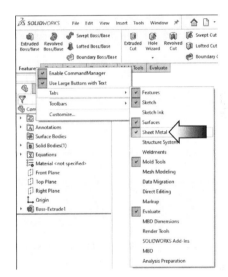

Right-click the **Evaluate** tab and select: **Tabs, Sheet Metal** tab (arrow).

Switch to the **Sheet Metal** tab.

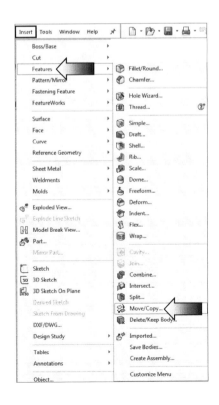

We will be creating a sheet metal part that fits over the existing model.

First, we need to make a copy of the original model and then convert it to sheet metal part.

3. Making a copy of the solid model:

Select **Insert / Features / Move/Copy** (arrow). If the Move/Copy options are not available, click the **Translate/Rotate** button at the bottom of the tree change to the **Move/Copy** options.

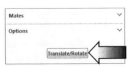

For Bodies to Move/Copy, <u>select the model</u> in the graphics area.

Enable the **Copy** checkbox (arrow).

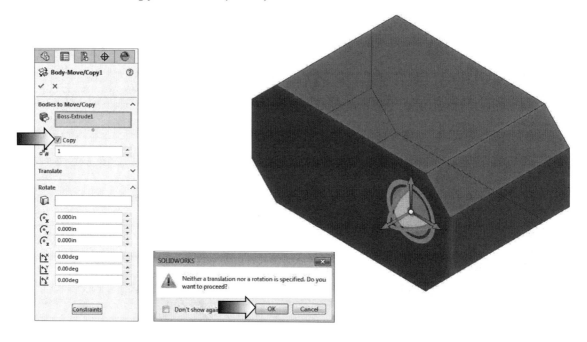

Keep all distances at their **Zero** values.

Click **OK**. Click OK again to proceed with the move.

A copy of the solid model is created and placed on top of the original body.

<u>Do not move</u> the copy. This image is made to show two solid bodies in the model.

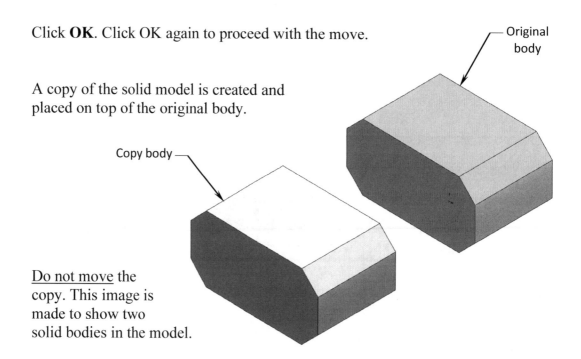

4. Converting to Sheet Metal:

Click **Convert to Sheet Metal**.

Select the **Fixed Face** as noted.

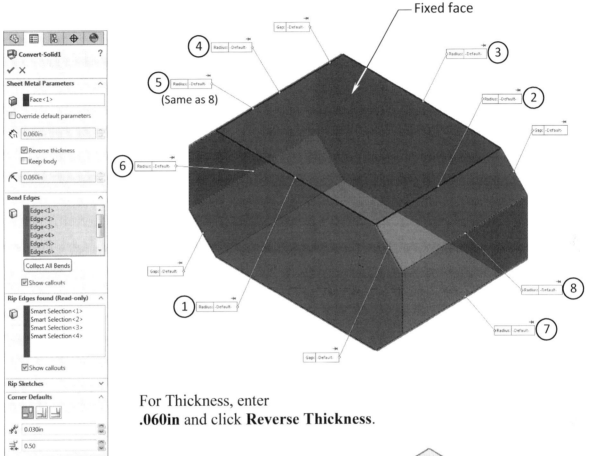

Fixed face

(Same as 8)

For Thickness, enter
.060in and click **Reverse Thickness**.

For Bend Edges, select
the **8 edges** as indicated.

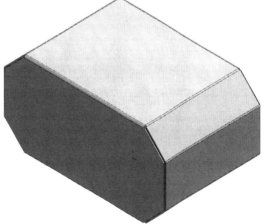

Enter **.030in**. for Gap.

Leave Overlap at the default **0.50**

Set Auto Relief to **Tear, 0.50**

Click **OK**.

5. Calculating the mass:

Change the **Material** to **1060 Alloy**.

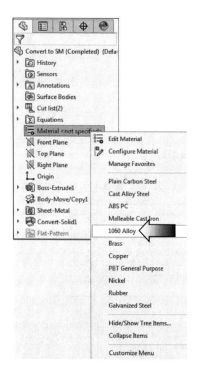

Switch to the **Evaluate** tab.

Click **Mass Properties** .

Select the new **solid body**
(Convert Solid1) under the
Cut List folder.

Enter the mass of the
converted part here:

_____ lbs.

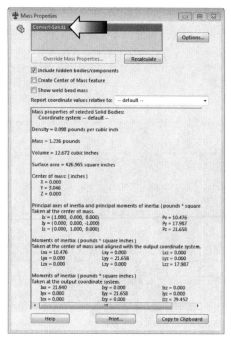

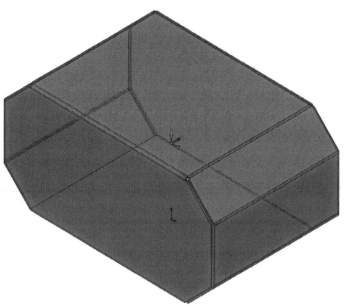

6. Measuring the Flat Pattern:

Right-click the converted model and select **Toggle Flat Display** (arrow).

This option "superimposes" the flat pattern over the folded for viewing purposes.

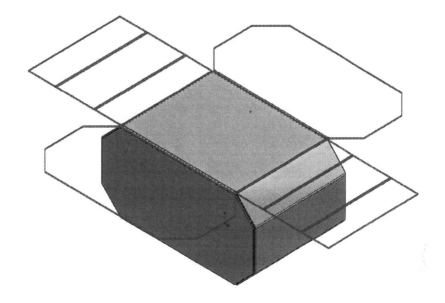

Press **Esc** to cancel the Toggle Flat Pattern mode.

Click the **Flatten** command on the Sheet Metal tab to flatten the part.

Switch to the **Evaluate** tab.

Click **Measure**.

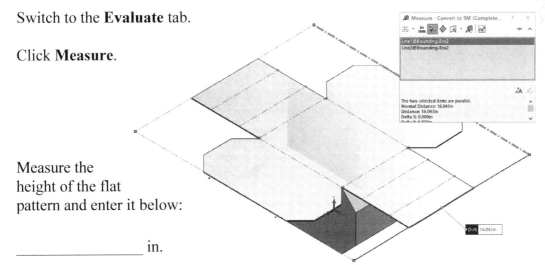

Measure the height of the flat pattern and enter it below:

_____ in.

7. Creating a Hem:

Turn-off the Flatten command and click **Hem** .

Select the **edge** as indicated and select the **Rolled** type (arrow).

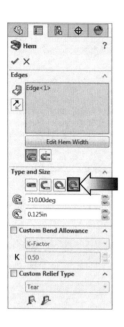

Enter **310deg** for Angle.

Enter **.125in** for Radius.

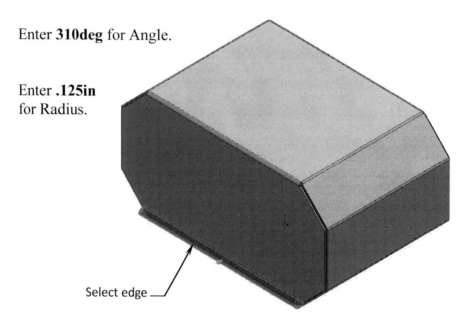

Select edge —

Click **Edit Hem Width** (arrow).

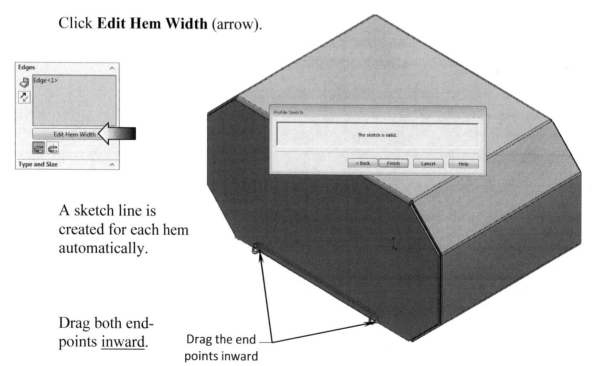

A sketch line is created for each hem automatically.

Drag both end-points <u>inward</u>.

Drag the end points inward

The line has an On-Edge relation with the edge of the model, but the 2 ends should be located with distance dimensions.

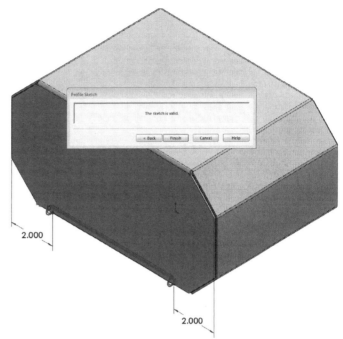

Add a distance of **2.000in** to each end of the line as shown.

Attach the dimensions to the outer edges of the flange.

Click **Finish**.

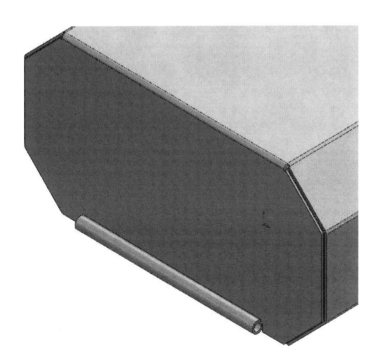

Isometric View

Right View

8. Calculating the mass:

Switch to the **Evaluate** tab.

Click **Mass Properties** 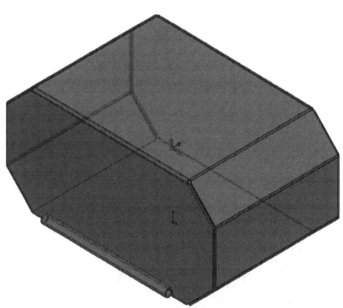.

NOTE: It is important when selecting the solid body to calculate the mass. Be sure to select the converted body every time.

Expand the **Cut List** and the **Sheet Metal** folders.

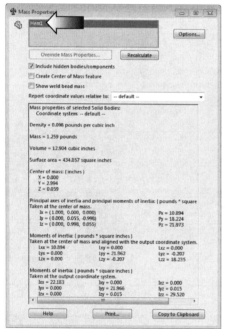

Select the converted model (Hem1) and enter the mass here:

_____ lbs.

Save and close all documents.

CHALLENGE 2

1. Opening a part document:

Select **File / Open**.

Browse to the Training Folder and open the part document named **SM Cut List.sldprt**.

This challenge examines your skills on modifying the bend radiuses and retrieving the Cut List Properties.

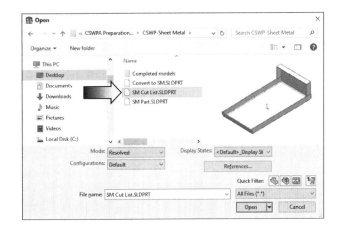

2. Modifying the first bend radius:

This model has several flanges, but we will only change the radius dimensions of the Edge Flange1 and Edge Flange2.

Locate the **Edge Flange1** from the FeatureManager tree.
Right-click it and select **Edit Feature** (arrow).

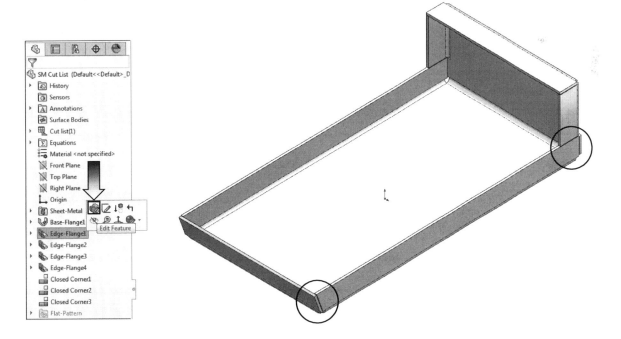

Enable the **Use Gauge Table**
checkbox.

Change the default bend
radius to **.125in** (arrow).

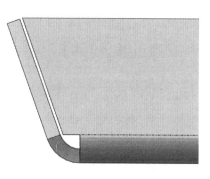

Keep all other parameters
at their default values.

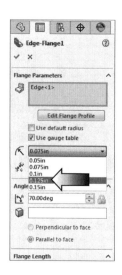

Click **OK**.

Click **Flatten**.

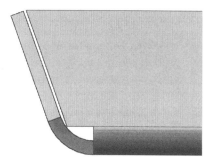

3. Measuring the flat length:

Switch to the **Evaluate** tab.

Click **Measure**.

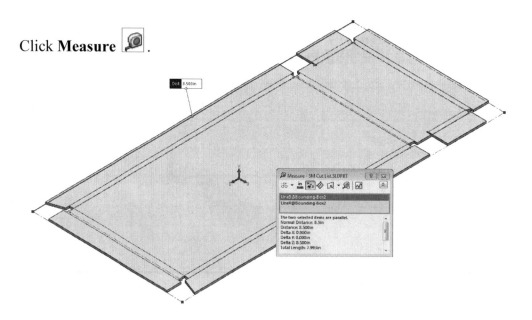

Measure the <u>length</u> of the flat pattern (the longest length) and enter it below:

_____ in.

4. Modifying the 2nd bend radius:

Click off the Flatten command. Locate **Edge Flange2** from the FeatureManager, right-click it and select **Edit Feature** (arrow).

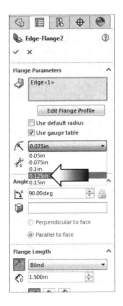
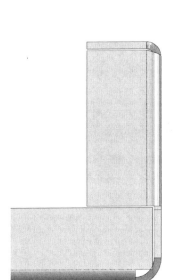

Enable the **Use Gauge Table** checkbox.

Change the bend radius to **.125in**

Keep all other parameters at their default values.

Click **OK** and click **Flatten**.

5. Measuring the flat length:

Click **Measure** 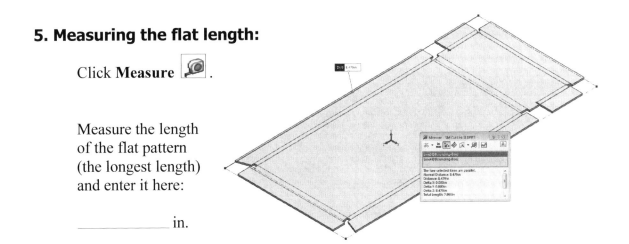.

Measure the length of the flat pattern (the longest length) and enter it here:

_____ in.

6. Finding the Bounding Box Area:

Right-click the **Cut list** 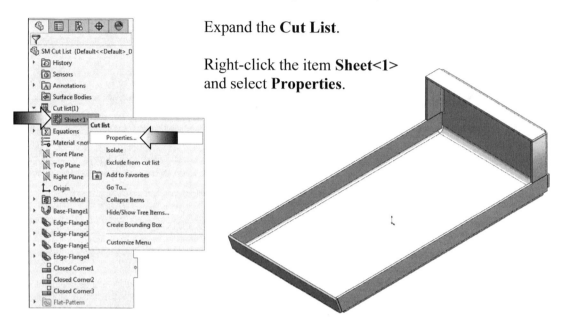 and select **Update** (arrow).

Expand the **Cut List**.

Right-click the item **Sheet<1>** and select **Properties**.

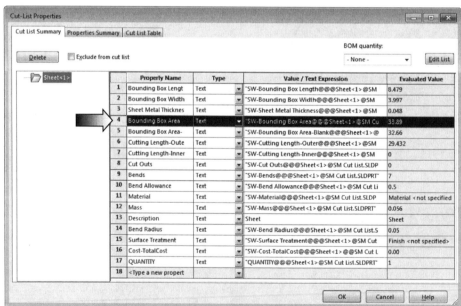

Locate row number 4 and enter the <u>Bounding Box Area</u> Evaluated Value

here: _____ in^2.

Save and close all documents.

CHALLENGE 3

1. Opening a part document:

Select **File / Open**.

Browse to the Training Folder
and open a part document
named **SM Part.sldprt**.

This challenge reinforces your
skills on creating a sheet metal
part and using the forming tools.

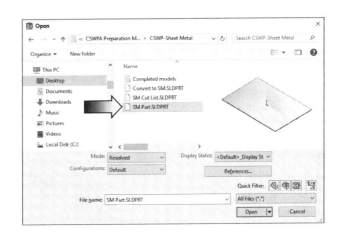

2. Creating the Edge Flanges:

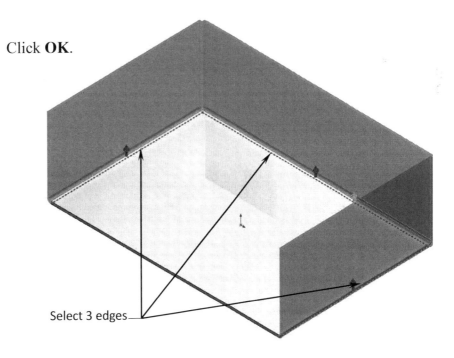

Click the **Edge Flange** command.

Select the **3 edges** of the Base Flange as noted.

Set the following: *** Gap: .010in**   *** Blind: 2.000in**

         * Outer Virtual Sharp   *** Bend Outside**

Click **OK**.

Select 3 edges

3. Closing the corners:

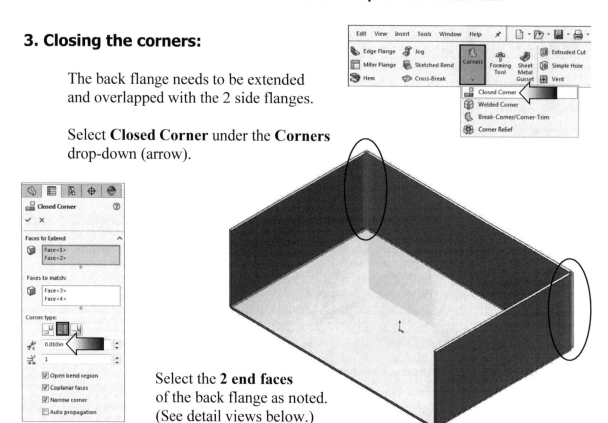

The back flange needs to be extended and overlapped with the 2 side flanges.

Select **Closed Corner** under the **Corners** drop-down (arrow).

Select the **2 end faces** of the back flange as noted. (See detail views below.)

Select the **Overlap** option and enter **.010** for Gap Distance (arrow).

Keep the Overlap/Underlap Ratio at **1** (full material thickness).

Enable **Open Bend Region**, **Coplanar Faces**, and **Narrow Corner** checkboxes.

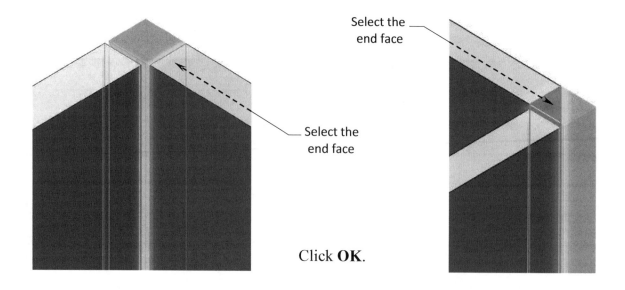

Select the end face

Select the end face

Click **OK**.

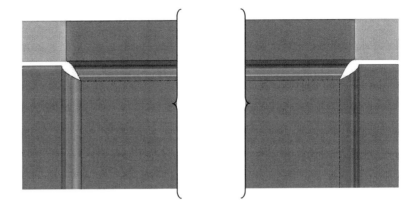

Change to the Top
View (Control+5)
and zoom in on the
2 corners to verify
the Closed Corner
features.

4. Creating a cutout:

Select the sketch face as noted and open a **new sketch**.

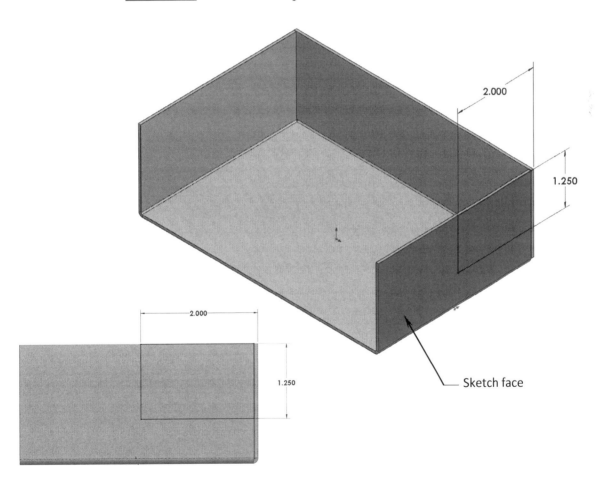

Sketch a **Rectangle** and add the height and width dimensions as shown.
(The dimension 2.00in is measured to the outer edge of the back flange.)

Switch to the **Features** tab and click **Extruded Cut** .

Use the default **Blind** cut and enable the **Link to Thickness** checkbox.

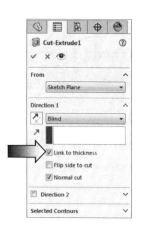

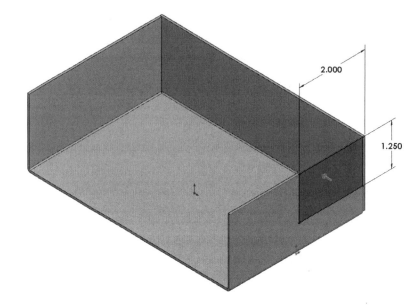

Click **OK**.

5. Adding a Miter Flange:

Select the <u>top face</u> of the back flange and open a **new sketch**.

Sketch **2 lines** from the upper right corner of the flange.

Add the dimensions shown to fully define the sketch.

Click the **Miter Flange** command.

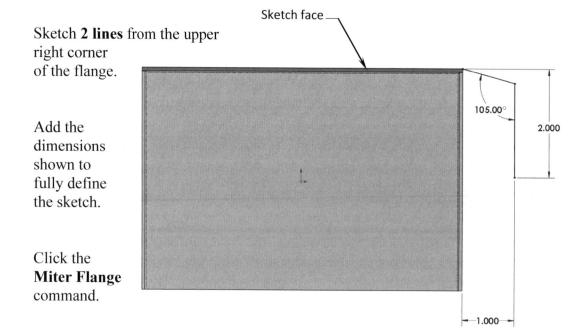

Sketch face

105.00°

2.000

1.000

Set the following to define the Miter Flange:

* **Use Default Radius**   * **Flange Position: Material Inside**

* **Gap Distance: .010in**   * **End Offset: .720in**

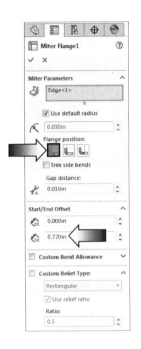

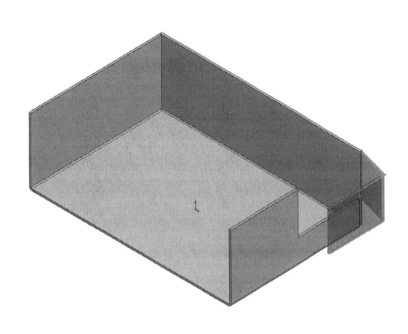

NOTE: *The dimension End Offset .720in (arrow) creates the required gap **.060in** for this challenge. This .060in gap must remain the same from beginning to end.*

Change to the Right view orientation (Control+4).

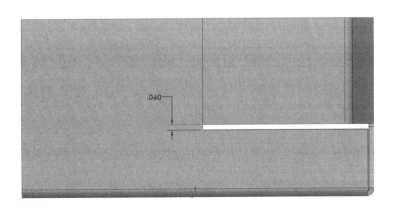

Use the **Measure** tool to inspect the gap.

6. Adding a Miter Flange:

Select the <u>edge</u> as noted and click the **Edge Flange** command.

Move the cursor inward and click the mouse to lock the direction.

Select the **Bend Outside** option under the Flange Position section.

Click the **Edit-Flange Profile** button to edit the sketch.

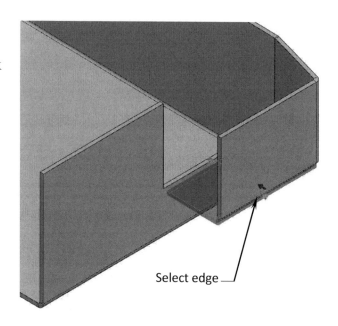

Select edge

Create an offset of **.060in** from the 2 edges indicated in the image.

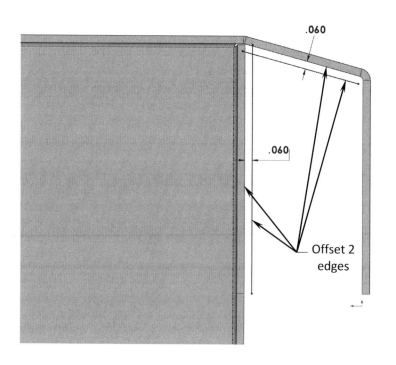

.060

.060

Offset 2 edges

Click **Reverse** offset to place the lines inside.

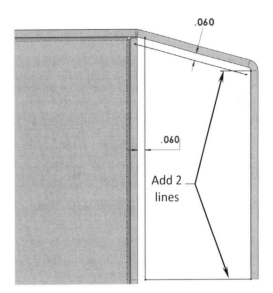

Sketch **2** additional **Lines** as noted. (Lock the right ends of the lines to the existing vertices with the coincident relations.)

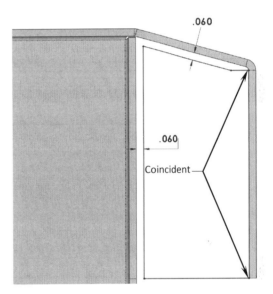

Trim the overlaps to create a closed profile.

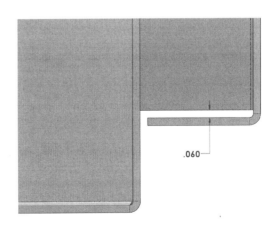

Change to the **Front** orientation (Control+1) and verify the gap between the new flange and the part. The gap should be .060in.

Click **Finish**.

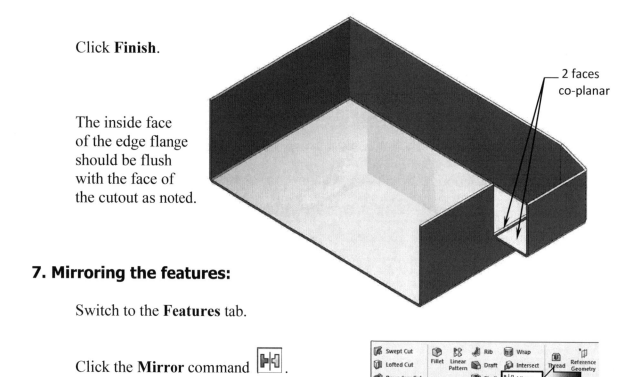

2 faces co-planar

The inside face of the edge flange should be flush with the face of the cutout as noted.

7. Mirroring the features:

Switch to the **Features** tab.

Click the **Mirror** command.

For Mirror Face/Plane, select the **Right** plane from the Feature tree.

For Features to Mirror, select the **Cut-Extrude1**, the **Miter Flange1**, and the **Edge-Flange1** from the FeatureManager tree.

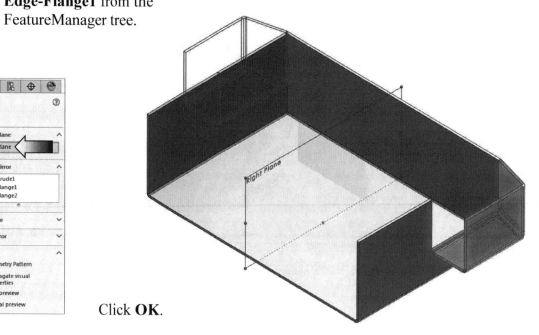

Click **OK**.

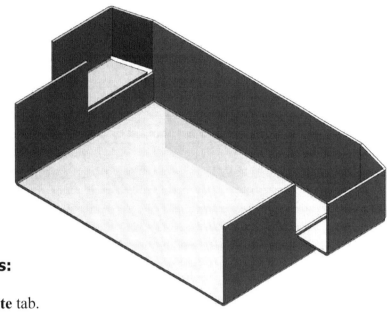

The selected features are mirrored about the Right plane.

The material **1060 Alloy** has already been assigned to this model.

8. Calculating the mass:

Switch to the **Evaluate** tab.

Click **Mass Properties** 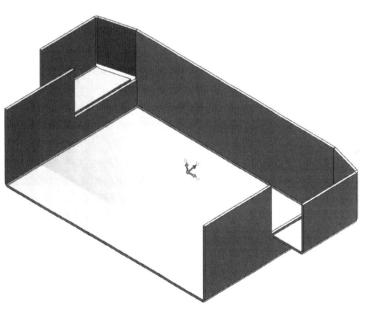.

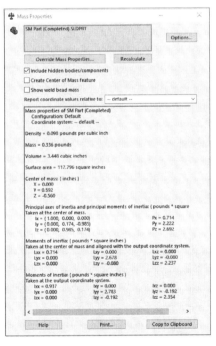

Locate the mass of the part and enter it below:

_____ lbs.

9. Changing the K-Factor value:

Click the **Sheet-Metal** folder and select **Edit Feature** (arrow).

Change the **K-Factor** to **.48**

Leave other parameters at their default values.

Click **OK**.

10. Measuring the flat length:

Select the **Measure** command 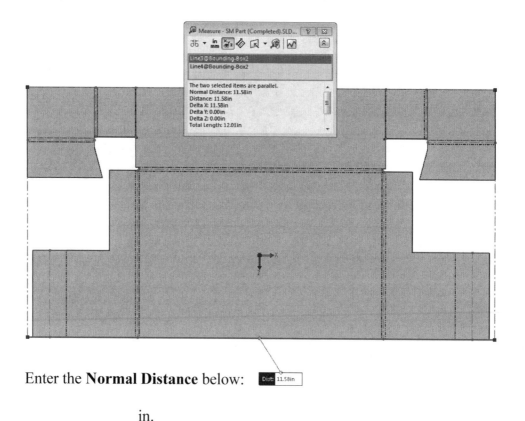 from the **Evaluate** tab.

Measure the **overall width** dimension (the longest length) of the part.

Enter the **Normal Distance** below: Dist: 11.58in

_____ in.

11. Adding Hems:

Click the **Hem** command .

Select the **2 outer edges** as indicated.

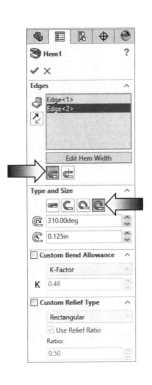

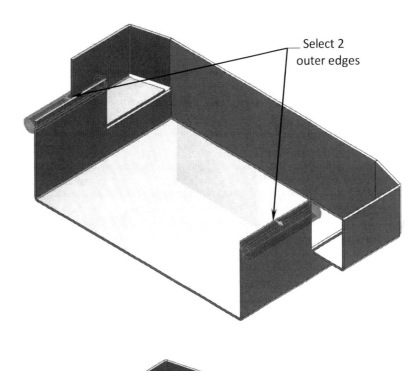

Select 2 outer edges

Set the material to **Inside** (arrow).

For Type, select **Tear Drop**.

Set Angle to **310deg**.

Set Radius to **.125in**.

Keep all other parameters at their default values.

Click **OK**.

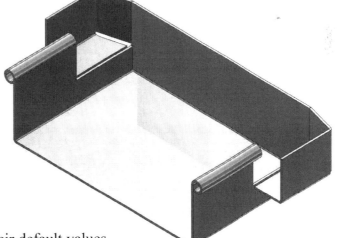

12. Adding a form feature:

Expand the **Design Library**, the **Forming Tools**, and the **Louvers folders**.

Drag and drop the **Louver** form tool to the <u>inside face</u> of the part as shown.

Change the Rotation Angle to **270deg**.

Click **Flip Tool** to flip to louver outward.

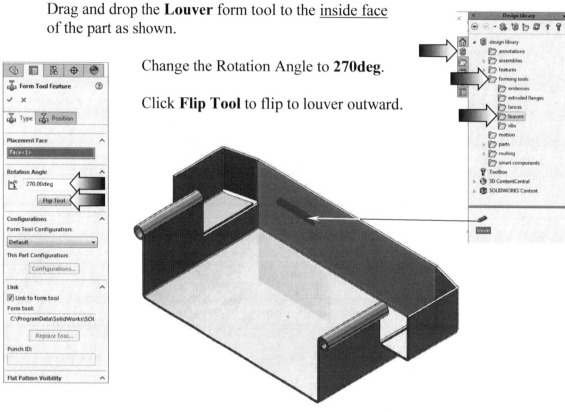

Click the **Position** tab (arrow).

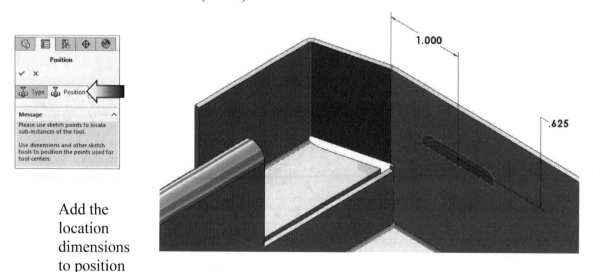

Add the location dimensions to position the louver and click **Finish**.

13. Creating a linear pattern:

Switch to the **Features** tab and select the **Linear Pattern** command (arrow).

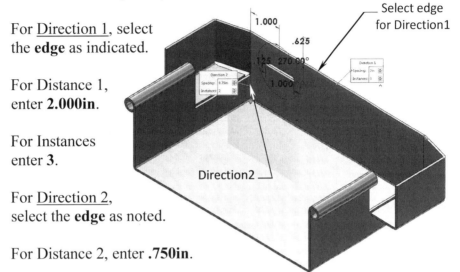

For <u>Direction 1</u>, select the **edge** as indicated.

For Distance 1, enter **2.000in**.

For Instances enter **3**.

For <u>Direction 2</u>, select the **edge** as noted.

For Distance 2, enter **.750in**.

For Instances, enter **2**.

For Features to Pattern, select the **Louver**.

Click **OK**.

14. Calculating the mass:

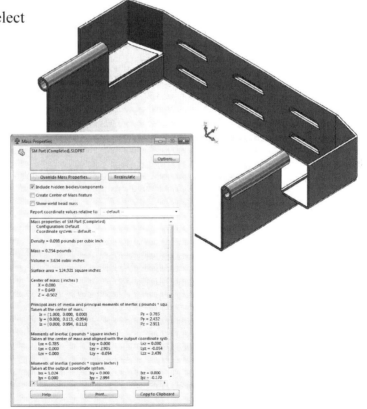

Switch to the **Evaluate** tab and click **Mass Properties** ⚖️ .

Locate the mass of the part and enter it below:

_____ lbs.

Save and close.

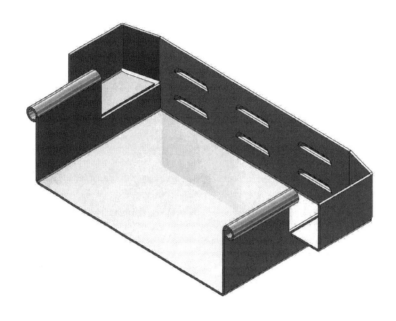

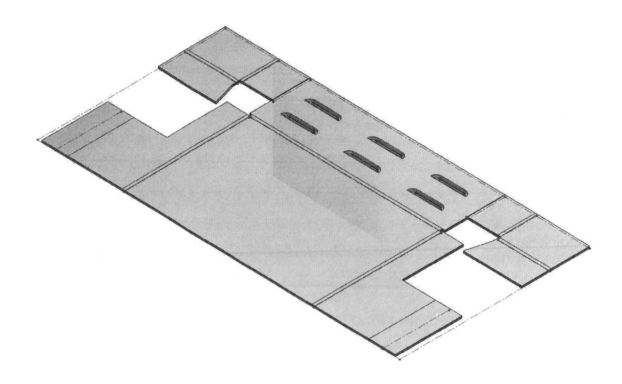

CHAPTER 5

CSWP – Advanced Surfacing

CSWP – Advanced Surfacing

The completion of the Certified SOLIDWORKS Professional Advanced Surfacing (CSWPA-SU) exam proves that you have successfully demonstrated your ability to use SOLIDWORKS Advanced Surfacing tools.

Successful completion of this exam also shows the ability to create advanced surface models as well as troubleshoot and fix broken surface bodies or imported bodies that are incorrect using advanced surfacing techniques.

Recommended Training Courses: Advanced Surface Modeling.

Note: You must use at least SOLIDWORKS 2017 for this exam. Any use of a previous version will result in the inability to open some of the testing files.

Exam Length: 90 minutes
Minimum Passing grade: 75%
Re-test Policy: There is a minimum 14 days waiting period between every attempt of the CSWPA-SU exam. Also, a CSWPA-SU exam credit must be purchased for each exam attempt.

All candidates receive electronic certificates and personal listing on the CSWP directory when they pass.

Exam features hands-on challenges in many of these areas of the SOLIDWORKS Surfacing functionality:

Splines, 3D Curves, Boundary Surface, Filled Surface, Swept Surface, Planar Surface, Knit Surface, Trim Surface, Untrim Surface, Move Face, Extend-Surface, Fillets, and thicken.

CSWP – Advanced Surfacing

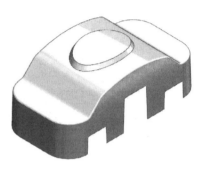

View Orientation Hot Keys:

Ctrl + 1 = Front View
Ctrl + 2 = Back View
Ctrl + 3 = Left View
Ctrl + 4 = Right View
Ctrl + 5 = Top View
Ctrl + 6 = Bottom View
Ctrl + 7 = Isometric View
Ctrl + 8 = Normal To Selection

Dimensioning Standards: **ANSI**

Units: **INCHES** – 3 Decimals

Tools Needed:

Extruded Surface	Planar Surface	Filled Surface
Move/Copy	Knit Surface	Offset Surface
Trim Surface	Cut with Surface	Mass Properties

CHALLENGE 1

1. Opening a part document:

Select **File / Open**.

Browse to the Training Folder and open a part document named:
Surface Patch.sldprt.

This challenge examines your skills on using the options in the Filled Surfaces, Swept Surfaces, Delete Surfaces and Knit Surfaces.

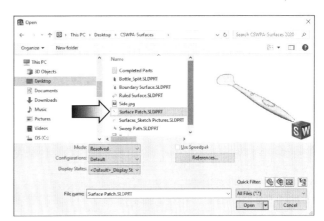

2. Enabling the Surfaces tab:

Right-click the **Evaluate** tab and enable the **Surfaces** tab (arrow).

There are two openings in this surface model.

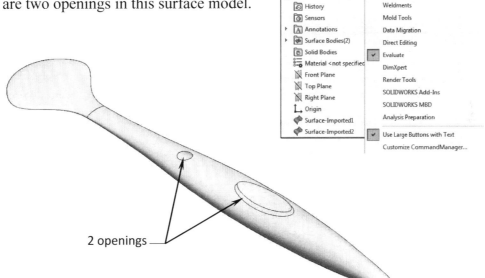

2 openings

You are required to patch up the elliptical opening using any of the surface tools, and remove the circular hole completely without leaving any trace of it.

3. Patching the 1st opening:

Switch to the **Surfaces** tab.

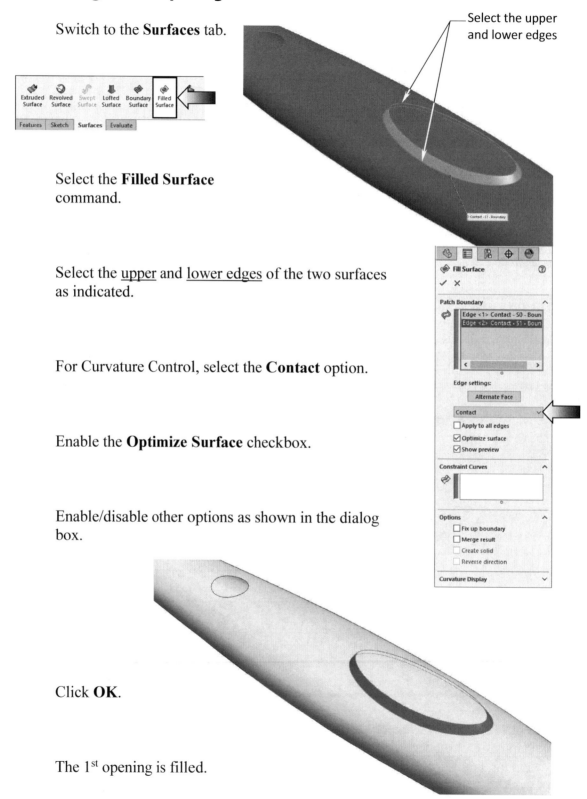

Select the upper
and lower edges

Select the **Filled Surface**
command.

Select the <u>upper</u> and <u>lower edges</u> of the two surfaces
as indicated.

For Curvature Control, select the **Contact** option.

Enable the **Optimize Surface** checkbox.

Enable/disable other options as shown in the dialog
box.

Click **OK**.

The 1st opening is filled.

4. Measuring the surface area:

Switch to the **Evaluate** tab.

Select the **Measure** command.

Click the surface of the patch and enter the **Area** below: (round-off to 2 decimals)

_____ in^2.

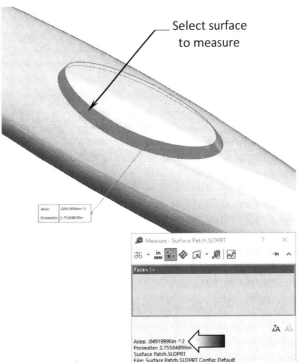

Select surface to measure

5. Patching the 2nd opening:

Next, we will delete and patch the circular opening.

Select the <u>edge</u> of the circular hole.

Press the **Delete** key on the <u>keyboard</u>.

Select the **Delete Hole** option (arrow).

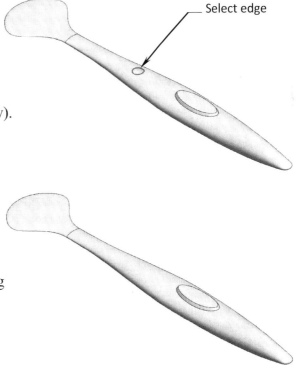

Select edge

The hole is removed and the opening is patched automatically.

Click **OK**.

6. Knitting the surfaces:

Select the **Knit Surface** command from the **Surfaces** tab.

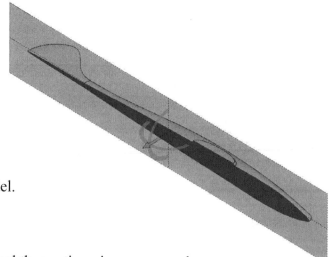

For Surfaces to Knit, select 3 surfaces.

Select 3 surfaces

Enable the check-boxes: **Create Solid** and **Merge Entities**.

Click **OK**.

7. Creating a section view:

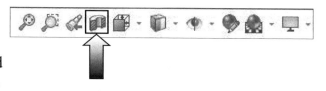

Click the **Section View** command from the View Heads-Up toolbar.

Use the **Front** plane as the Cutting plane.

The Blue color represents the solid material.

The surface model has been thickened into a solid model.

Push **Esc** to cancel the section view command.

8. Calculating the mass:

Right-click the **Material** option and select:
Nickel (arrow).

Switch to the **Evaluate** tab.

Click **Mass Properties** .

Locate the mass of the part and enter it below:
(round-off to 2 decimal places)

_____ lbs.

9. Saving your work:

Save your work as **Surface Patch (Completed).sldprt**.

CHALLENGE 2

1. Opening a part document:

Select **File / Open**.

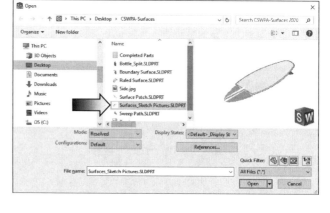

Browse to the Training Folder
and open a part document
named: **Surfaces_Sketch
Pictures.sldprt**.

This challenge will examine
your skills on working with
sketch pictures and using various surface tools to create the body of a surfboard.

2. Examining the sketch pictures:

There are 2 sketch pictures that have been previously inserted onto the Top and the
Right planes.

Expand Sketch1 and Sketch2 on the FeatureManager tree to see the two **Sketch
Pictures** (arrow).

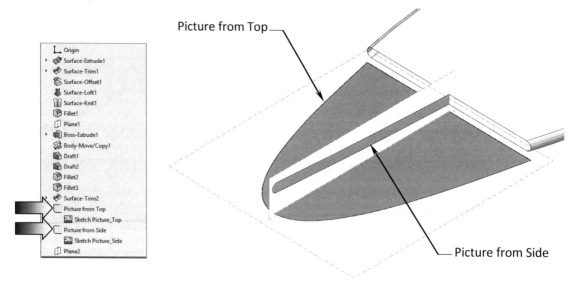

The Picture from Top represents the top profile of the surfboard, and the Picture
from Side represents the side profile.

3. Creating the 1st loft profile:

There will be a total of 3 sketches needed to create the tip-portion of the surfboard, two Loft Profiles and one Guide Curve.

Select the <u>Right</u> plane and open a **new sketch**.

Create a **2-Point Spline** along the edge as shown below.

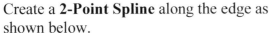

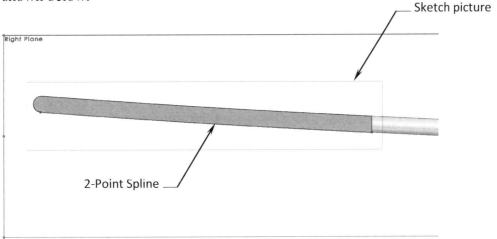

Sketch picture

Right Plane

2-Point Spline

Continue with tracing the outline of the picture using the **Tangent Arc**, **3-Point-Arc** and **Centerline** commands.

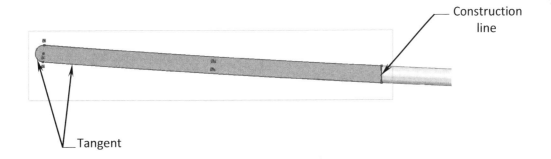

Construction line

Tangent

Add the relations as noted to fully define the sketch. This sketch will be the **Loft-Profile** number 1.

<u>Exit</u> the sketch or press Control+B.

4. Creating the 2nd loft profile:

The **Loft Profile** number 2 will be created next.

Select Plane2 and open a **new sketch**.

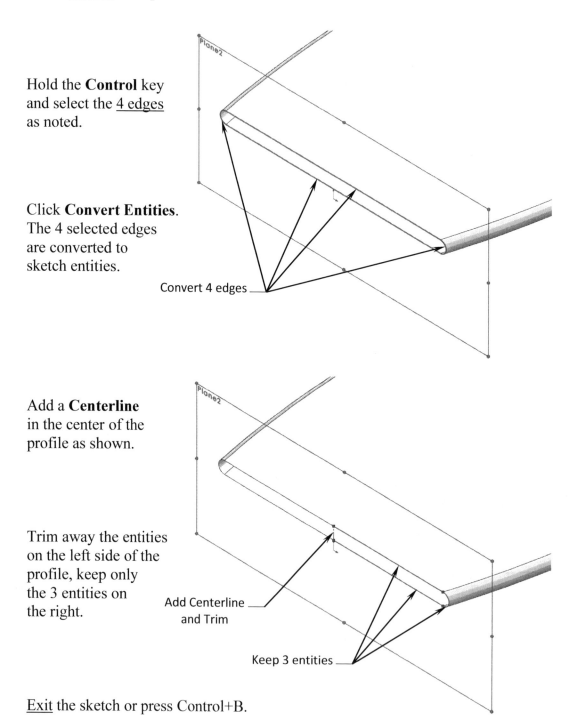

Hold the **Control** key and select the 4 edges as noted.

Click **Convert Entities**. The 4 selected edges are converted to sketch entities.

Convert 4 edges

Add a **Centerline** in the center of the profile as shown.

Trim away the entities on the left side of the profile, keep only the 3 entities on the right.

Add Centerline and Trim

Keep 3 entities

Exit the sketch or press Control+B.

5. Making the Guide Curve:

The guide curve will be created as a 3D Sketch. It is used to connect the 2 loft profiles and also as a guide curve when making the loft feature.

Select **3D Sketch** under the Sketch drop-down.

Sketch a **2-Point Spline** as shown. Make sure both ends of the spline are **Coincident** to the endpoints of the
2 loft profiles as indicated.

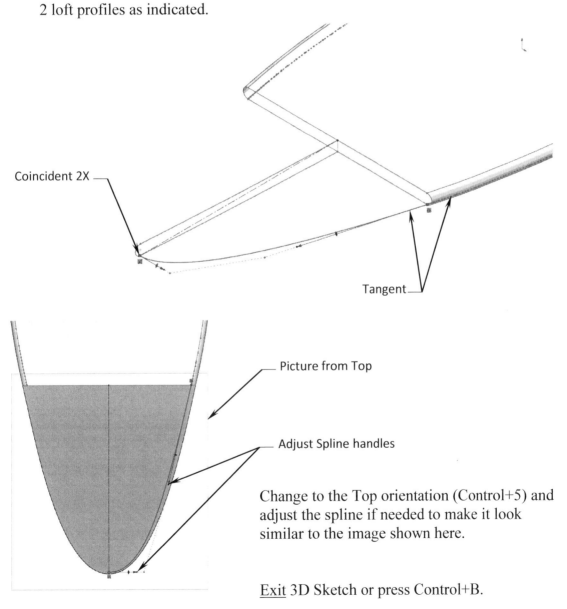

Coincident 2X

Tangent

Picture from Top

Adjust Spline handles

Change to the Top orientation (Control+5) and adjust the spline if needed to make it look similar to the image shown here.

Exit 3D Sketch or press Control+B.

6. Creating a lofted surface:

Switch to the **Surfaces** tab.

Select the **Lofted Surface** command .

For Loft Profiles, select the **Sketch4** and **Sketch5** as noted.

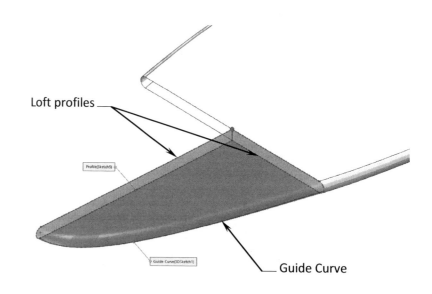

For Guide Curves, select the **3D Sketch** either from the FeatureManager tree or directly in graphics area.

Click **OK**.

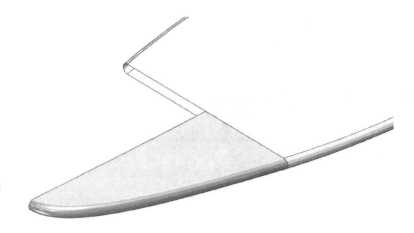

The left-half portion of the surfboard's tip is created.

7. Measuring the surface area:

Switch to the **Evaluate** tab.

Click **Measure** .

Select <u>only</u> the **Lofted Surface** from the graphics area.

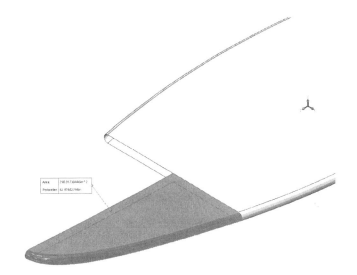

Locate the total surface **Area** and enter it below: (round-off to 2 decimal places).

_____ in ^2

8. Mirroring a surface body:

Switch to the **Features** tab.

Click **Mirror** .

For Mirror Face/Plane, select the **Right** plane from the FeatureManager tree.

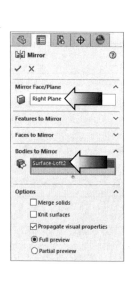 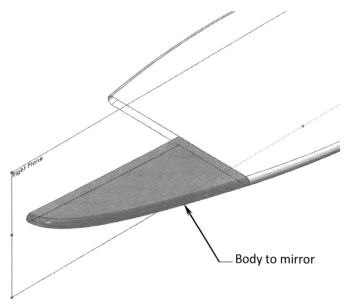

Body to mirror

Expand the <u>Bodies to Mirror</u> section and select the **Lofted Surface** as noted.

Select/deselect the checkboxes shown in the dialog box.

Click **OK**.

Inspect the surfaces from different angles to ensure they blended correctly, all around.

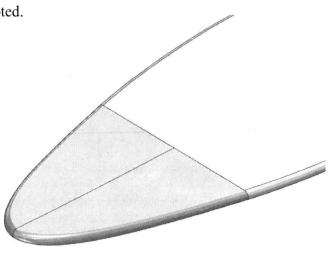

9. Knitting the surfaces:

Switch to the **Surfaces** tab and click **Knit Surface**.

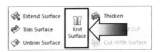

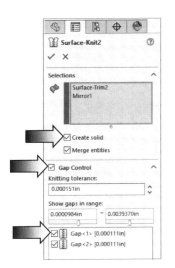

For Surfaces to Knit, select 3 surfaces as noted.

Enable: **Create Solid**, **Merge Entities**, and **Gap Control**.

Enable the checkboxes next to Gaps.

Click **OK**.

Select 3 surfaces

10. Combining the solid bodies:

The surfboard body and the 2 fins must be combined into a single body before the final mass can be calculated.

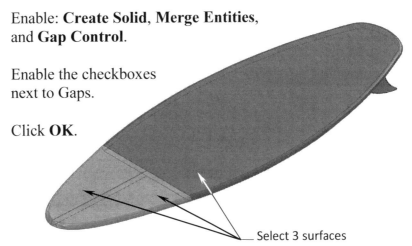

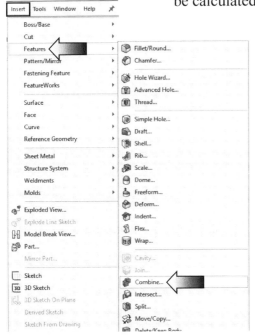

Select: **Insert**, **Features**, **Combine**.

Click the **Add** option and select the 3 Solid Bodies in the Solid Bodies folder.

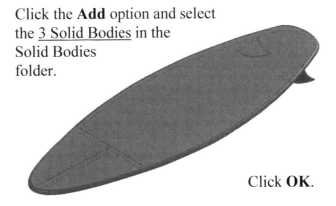

Click **OK**.

11. Assigning material:

Right-click the **Material** option on the Feature-Manager tree and select: **Edit Material**.

Expand the folder: **Other Non-Metals**

Select the material: **Polyurethane Foam Rigid**.

Click **Apply** and **Close**.

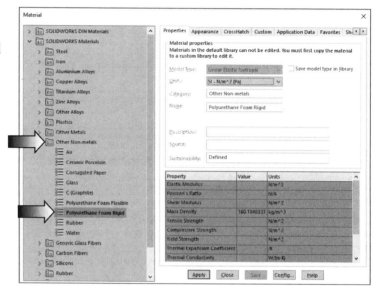

12. Calculating the final mass:

Switch to the **Evaluate** tab and click **Mass Properties** .

Locate the mass of the part and enter it below: (round-off to 2 decimal places)

_____ lbs.

13. Saving your work:

<u>**Save**</u> your work as **Surface Sketch Pictures (Completed).sldprt**.

CHALLENGE 3

1. Opening a part document:

Select **File / Open**.

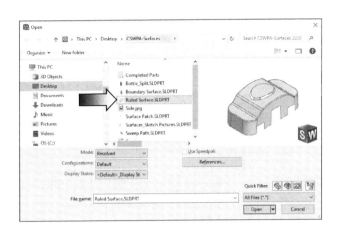

Browse to the Training Folder and open a part document named **Ruled Surface.sldprt**.

This challenge examines your skills on the use of some of the surfacing tools such as Ruled, Trim, and Knit Surface, etc.

2. Creating a Ruled Surface:

This model has 2 surface bodies. The gap between the 2 surfaces must be filled at 32° angle. One way to achieve this is to create a Ruled Surface with Tapered-to-Vector option.

Switch to the **Surfaces** tab and click **Ruled Surface**.

For Type, select the option: **Tapered to Vector**.
For Distance, enter **.600in**.

For Direction, select the **Top** plane.

For Angle, enter **32.00deg**.

For Edge Selection select the **edge** of the <u>upper surface</u>.

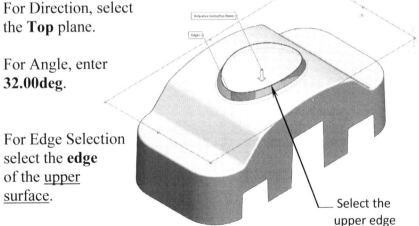

Select the upper edge

Click **OK**.

Rotate to the bottom orientation to inspect the Ruled Surface.

3. Trimming the surface bodies:

Select the **Trim Surface** command from the **Surfaces** tab.

For Trim Type, select **Mutual**.

Select 2 surfaces

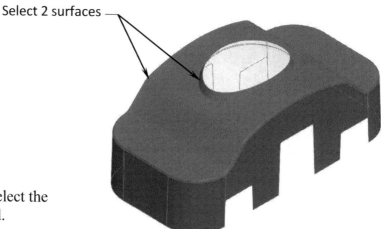

For Trim Selections, select the
2 surfaces as indicated.

Click the **Keep Selection** option (arrow)

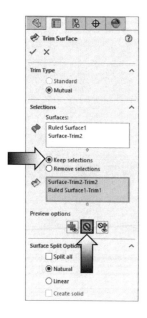

For clarity, click **Show Excluded Surfaces** under Preview
Options.

Select the **2 surfaces** to keep
and remove the extra
surface as shown
in the image.

Leave all other
options at their
defaults.

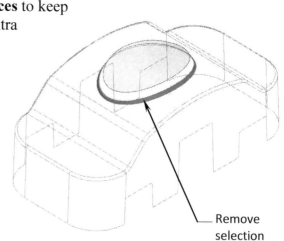

Remove
selection

Click **OK**.

4. Measuring the surface Area:

Switch to the **Evaluate** tab.

Click **Measure**  .

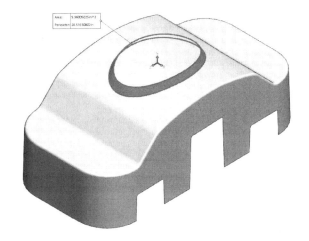

Select the **Ruled Surface** in the graphics area.

Locate the total surface Area and enter it below: (round-off to 2 decimal places).

_____ in ^2.

OPTIONAL:

Knit and thicken the surface model, use a wall thickness of .125in.

Create a section view to verify the thickness of the solid model.

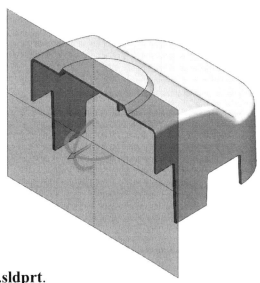

Save your work as **Ruled Surface (Completed).sldprt**.

CHALLENGE 4

1. Opening a part document:

Select **File / Open**.

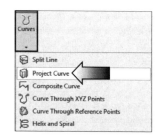

Browse to the Training Folder and open a part document named **Project Curve.sldprt**.

This challenge examines your skills on the use of the Project-Curve and Sweep with Guide Curves.

2. Creating a Projected Curve:

The option Project Curve/Sketch-to-Sketch creates a curve that represents the intersection of sketches from two intersecting planes.

Switch to the **Surfaces** tab and click: **Curves, Project Curve**.

For Projection Type, select: **Sketch on Sketch**.

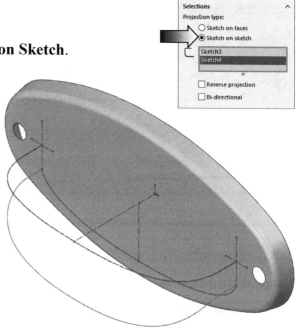

For Sketches to Project, select **Sketch3** and **Sketch4** from the FeatureManager tree.

The preview graphic shows a curve is being created from the intersection of the two sketches.

Click **OK**.

3. Sketching the sweep profile:

Select the <u>face</u> as indicated and open a **new sketch**.

Sketch an **Ellipse** as shown in the image.

Add a **Coincident** relation between the center of the ellipse and the end point of the path in Sketch3.

In order for the sweep feature to work properly, add the additional relations as indicated in the image below.

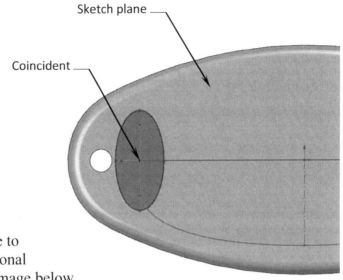

Add a **Coincident** between the left side of the ellipse and the left end of the construction line. Also add a **Pierce** relation between the bottom of the ellipse and the blue projected curve.

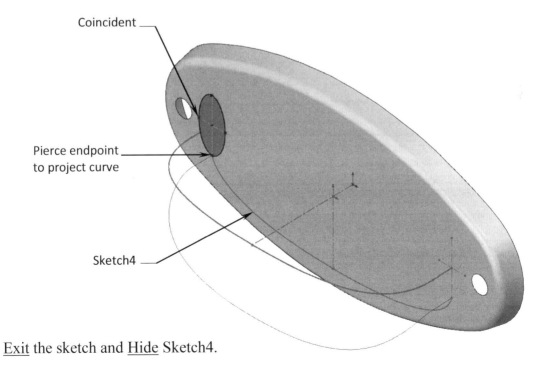

<u>Exit</u> the sketch and <u>Hide</u> Sketch4.

4. Creating a swept feature:

Click **Swept Surface**.

For Sweep Profile, select the **Ellipse** (Sketch5).

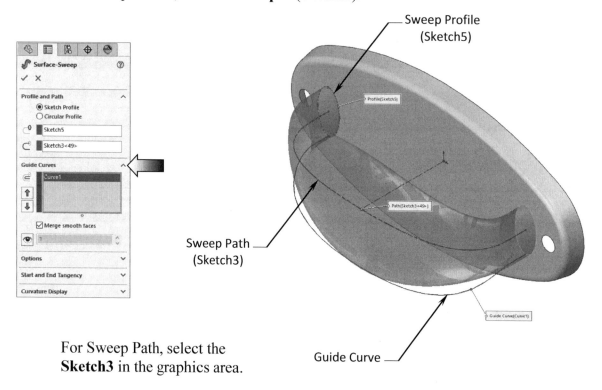

Sweep Profile
(Sketch5)

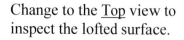

Sweep Path
(Sketch3)

Guide Curve

For Sweep Path, select the **Sketch3** in the graphics area.

Expand the Guide Curves section and select the **Projected Curve**.

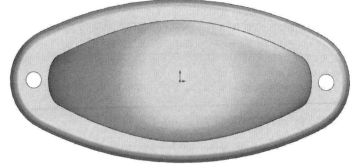

Click **OK**.

Change to the <u>Top</u> view to inspect the lofted surface.

(<u>Note:</u> There are slight differences between the SOLIDWORKS releases, so if the sweep fails, create a 2nd Ellipse and a 2nd Projected Curve, and use the Loft tool instead - 2 loft profiles, 2 guide curves, and 1 centerline parameter.)

5. Trimming the surfaces:

The bottom of the handle is closed off at this point. A trim feature is needed to open it up so that a thickness can be added to the model.

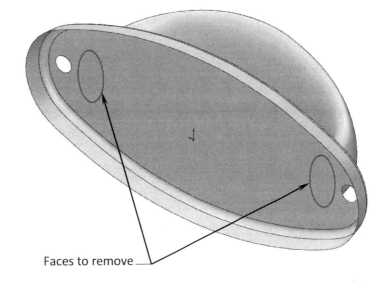

Click **Trim Surface**.

For Trim Type, select **Mutual**.

Faces to remove

For Selections, select the surfaces of the **Handle** and the **Base**.

For Remove Selection, select the **2 inner surfaces** of the handle as noted.

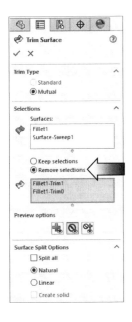

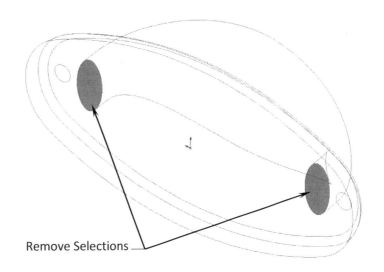

Remove Selections

Click **OK**.

6. Knitting the surfaces:

Click **Knit Surface**.

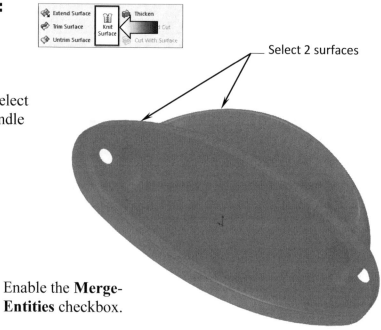

Select 2 surfaces

For Surfaces to Knit, select the **surfaces** of the Handle and the Base.

Enable the **Merge-Entities** checkbox.

Click **OK**.

7. Adding a .040in fillet:

Click **Fillet**.

Enter **.060in** for radius.

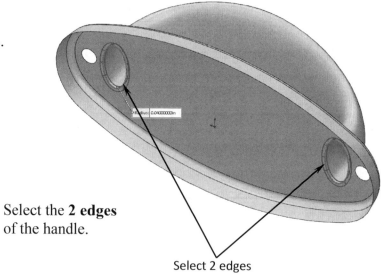

Select the **2 edges** of the handle.

Select 2 edges

Click **OK**.

8. Adding a thickness:

Click **Thicken**.

For Thicken Parameters, select the <u>surface model</u> in the graphics area.

For Thickness Direction, click **Inside**.

For Thickness, enter **.025in**.

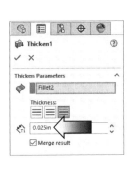

Add thickness
to inside

Click **OK**.

9. Calculating the mass:

Switch to the **Evaluate** tab and click **Mass Properties**.

The **Brass**
material has
already been
assigned to
the part.

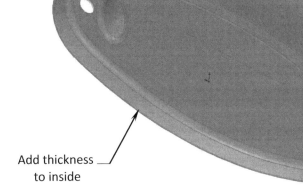

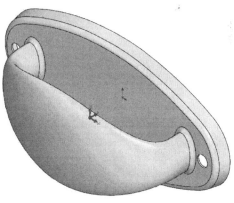

Locate the
mass of the
part and enter
it below:

_____ lbs.

<u>Save</u> your work as **Project Curve (completed).sldprt**.

CHALLENGE 5

1. Opening a part document:

Select **File / Open**.

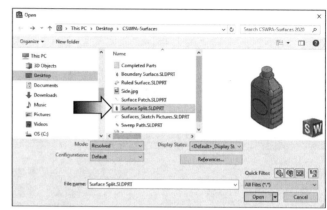

Browse to the Training Folder
and open a part document
named **Surface_Split.sldprt**.

This challenge examines your
skills on the use of the Split Line, Offset Surface and Lofted Surface tools.

2. Creating a split line:

The intent is to create a couple of indentations on
the front face of the bottle. There are several methods
to achieve this, but we will try using the Split Line option
for this challenge.

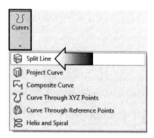

Click: **Curves**, **Split Line** on the **Surfaces** tab.

For Type of Split, select:
Projection

For Sketch to Project:
select the **Split Sketch** on
the FeatureManager tree.

For Faces to Split, select
the **front face** of the bottle.

Enable the boxes shown.

Click **OK**.

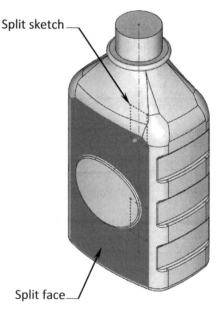

Split sketch

Split face

3. Creating the 1st offset surface:

Click **Offset Surface** .

For Offset Parameters, select the **2 surfaces** as indicated.

For Offset Distance, enter **.040in**. and click **Reverse**.

The new offset surfaces should be <u>inside</u> the model.

Click **OK**.

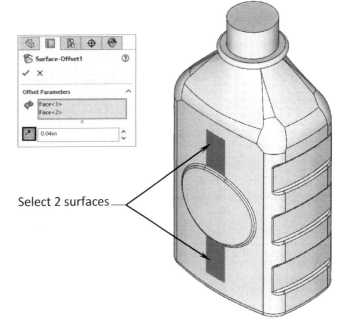

Select 2 surfaces

<u>Note:</u> *The offset distance .040in represents the depth of the recess feature that we are going to create in step number 5.*

4. Creating the 2nd offset surface:

Click **Offset Surface** again.

For Offset Parameters, select the **2 surfaces** as indicated.

For Offset Distance, enter **.040in**. and <u>clear</u> Reverse.

The new offset surfaces should be <u>outside</u> the model.

Click **OK**.

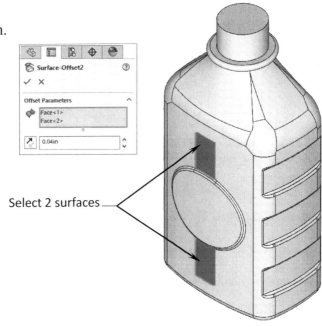

Select 2 surfaces

5. Creating the 1st lofted cut feature:

The two offset surfaces will be used to cut away the material in between them.

Switch to the **Features** bar.

Click **Lofted Cut**.

For Loft Profiles, select the **2 upper offset surfaces**.

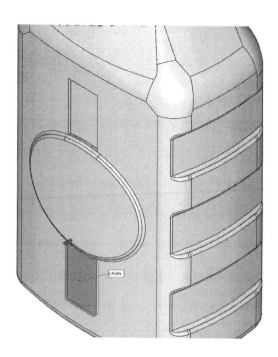

Drag to connectors to reposition them if needed. They should be positioned on top of one another.

Connectors

Click **OK**.

6. Creating the 2nd lofted cut feature:

Click **Lofted Cut** again .

For Loft Profiles, select the **2 lower offset surfaces**.

Make sure the two blue connectors are positioned correctly.

Click **OK**.

7. Adding the 1st .040in fillets:

Hide all 4 offset surfaces.

Click **Fillet** 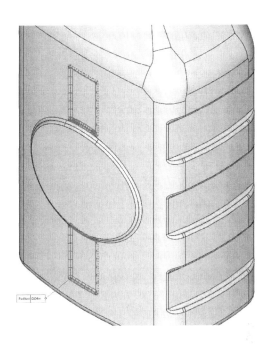.

Enter **.040in** for radius.

Select the 8 corner edges and all edges on the inside of the recessed features (select the faces also works well).

Click **OK**.

8. Adding the 2nd .040in fillets:

Click **Fillet** again.

For radius, enter **.040in**.

For edges to fillet, select all outside edges of the recessed features.

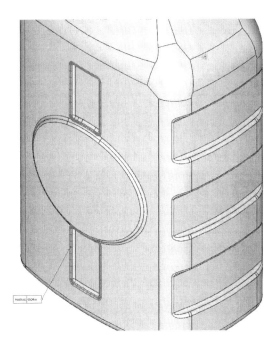

Click **OK**.

9. Shelling the solid body:

Click the **Shell** command from the **Features** tab.

Face to remove

For Wall Thickness, enter **.020in**.

For Faces to Remove, select the **top surface** of the bottle as noted.

Click **OK**.

10. Calculating the mass:

The material has already been selected as **PE High Density**.

Switch to the **Evaluate** tab.

Click **Mass Properties**

Locate the Mass of the model and enter it below: (Round-off to 2 decimals)

_____ lbs.

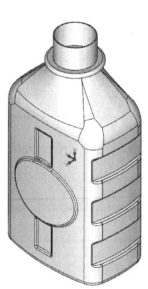

<u>Save</u> your work as:
Surface_Split (Completed).sldprt

Close all documents.

CHALLENGE 6

1. Opening a part document:

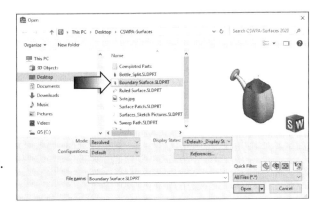

Select **File / Open**.

Browse to the Training Folder
and open a part document
named **Boundary Surface.sldprt**.

This challenge examines your
skills on the use of Splines, Planes,
Lofted Surface, Revolved Surface, Knit Surface and Boundary Surface.

2. Creating a lofted surface:

Switch to the **Surfaces** tab and click **Lofted Surface** .

For Loft Profiles, select **Sketch9** and **Sketch11** as
indicated.

For Guide Curves, select the **3 splines** using the
SelectionManager option.

SelectionManager

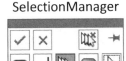

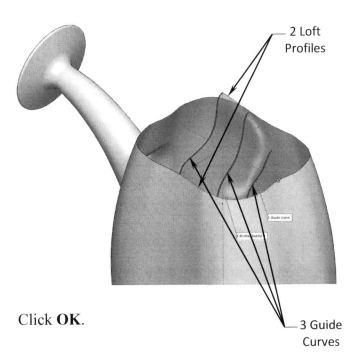

2 Loft
Profiles

3 Guide
Curves

Click **OK**.

3. Measuring the surface area:

Switch to the **Evaluate** tab.

Click **Measure**.

Select the **Surface Loft3** from the graphics area.

Enter the total **Area** below: (use 2 decimals)

_____ ^2

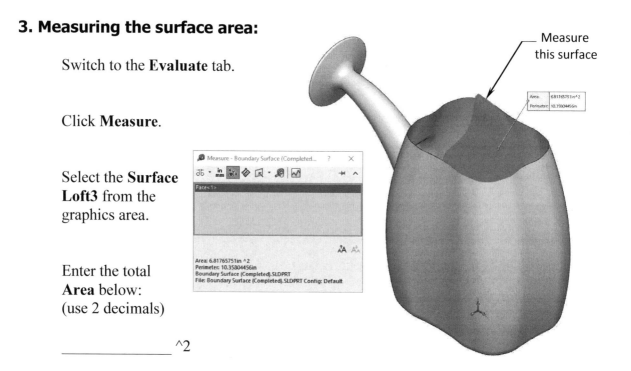

Measure this surface

4. Mirroring a surface body:

Switch to the **Features** tab and click **Mirror**.

For Mirror Face/Plane, select the **Right** plane from the FeatureManager tree.

For Bodies to Mirror select the **Surface-Loft3** from the graphics area.

Enable the **Knit-Surfaces** checkbox.

Click **OK**.

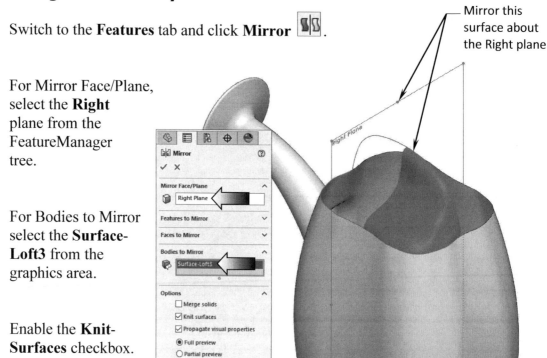

Mirror this surface about the Right plane

5. Mirroring a mirrored-surface-bodies:

Select the **Mirror** command again 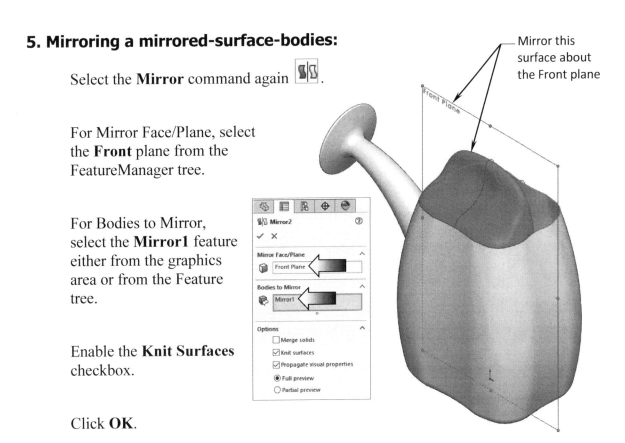.

For Mirror Face/Plane, select the **Front** plane from the FeatureManager tree.

For Bodies to Mirror, select the **Mirror1** feature either from the graphics area or from the Feature tree.

Enable the **Knit Surfaces** checkbox.

Click **OK**.

Mirror this surface about the Front plane

6. Creating the trimmed sketch:

Select the <u>Front</u> plane and open a **new sketch**.

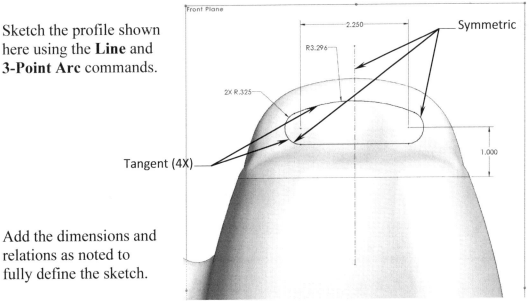

Sketch the profile shown here using the **Line** and **3-Point Arc** commands.

Add the dimensions and relations as noted to fully define the sketch.

7. Trimming the surface bodies:

Switch to the **Surfaces** tab.

Click **Trim Surface**.

For Trim Type, select **Standard**.

For Trim Tool, select **Sketch12**.

For Remove-Selections, the **2 inner portions** of the Sketch12, both front and back.

Click **OK**.

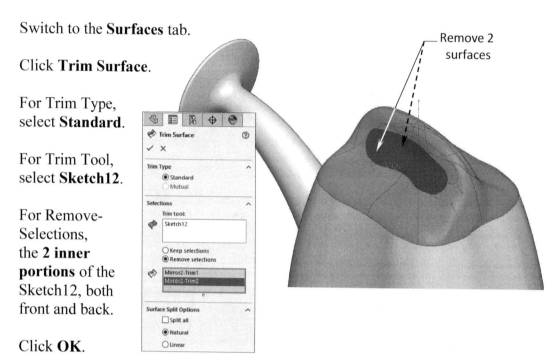

Remove 2 surfaces

8. Measuring the total area:

Switch to the **Evaluate** tab.

Click **Measure** 📷 .

Select the upper **surface** as noted.

Locate the total **Area** and enter it below: (Use 2 decimals).

_____ ^2

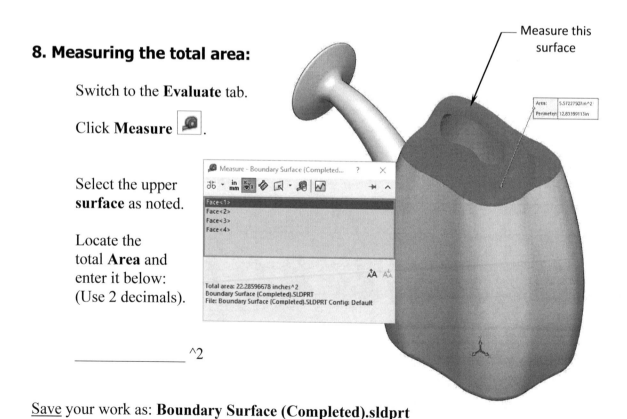

Measure this surface

Save your work as: **Boundary Surface (Completed).sldprt**

Glossary

Alloys:

An Alloy is a mixture of two or more metals (and sometimes a non-metal). The mixture is made by heating and melting the substances together.
Example of alloys are Bronze (Copper and Tin), Brass (Copper and Zinc), and Steel (Iron and Carbon).

Gravity and Mass:

Gravity is the force that pulls everything on earth toward the ground and makes things feel heavy. Gravity makes all falling bodies accelerate at a constant 32ft. per second (9.8 m/s). In the earth's atmosphere, air resistance slows acceleration. Only on airless Moon would a feather and a metal block fall to the ground together.
The mass of an object is the amount of material it contains.
A body with greater mass has more inertia; it needs a greater force to accelerate.
Weight depends on the force of gravity, but mass does not.

When an object spins around another (for example: a satellite orbiting the earth) it is pushed outward. Two forces are at work here: Centrifugal (pushing outward) and Centripetal (pulling inward). If you whirl a ball around you on a string, you pull it inward (Centripetal force). The ball seems to pull outward (Centrifugal force) and if released will fly off in a straight line.

Heat:

Heat is a form of energy and can move from one substance to another in one of three ways: by Convection, by Radiation, and by Conduction.

Convection takes place only in liquids like water (for example: water in a kettle) and gases (for example: air warmed by a heat source such as a fire or radiator).

When liquid or gas is heated, it expands and becomes less dense. Warm air above the radiator rises and cool air moves in to take its place, creating a convection current.

Radiation is the movement of heat through the air. Heat forms match set molecules of air moving and rays of heat spread out around the heat source.

Conduction occurs in solids such as metals. The handle of a metal spoon left in boiling liquid warms up as molecules at the heated end moves faster and collide with their neighbors, setting them moving. The heat travels through the metal, which is a good conductor of heat.

Inertia:

A body with a large mass is harder to start and also to stop. A heavy truck traveling at 50mph needs more power breaks to stop its motion than a smaller car traveling at the same speed.

Inertia is the tendency of an object either to stay still or to move steadily in a straight line, unless another force (such as a brick wall stopping the vehicle) makes it behave differently.

Joules:

Joules is the SI unit of work or energy.

One Joule of work is done when a force of one Newton moves through a distance of one meter. The Joule is named after the English scientist James Joule (1818-1889).

Materials:

Stainless steel is an alloy made of steel with chromium or nickel.

Steel is made by the basic oxygen process. The raw material is about three parts melted iron and one part scrap steel. Blowing oxygen into the melted iron raises the temperature and gets rid of impurities.

All plastics are chemical compounds called polymers.

Glass is made by mixing and heating sand, limestone, and soda ash. When these ingredients melt they turn into glass, which is hardened when it cools.

Glass is in fact not a solid but a "supercooled" liquid; it can be shaped by blowing, pressing, drawing, casting into molds, rolling, and floating across molten tin, to make large sheets.

Ceramic objects, such as pottery and porcelain, electrical insulators, bricks, and roof tiles are all made from clay. The clay is shaped or molded when wet and soft, and heated in a kiln until it hardens.

Machine Tools:

Are powered tools used for shaping metal or other materials, by drilling holes, chiseling, grinding, pressing, or cutting. Often the material (the work piece) is moved while the tool stays still (lathe), or vice versa, the work piece stays while the tool moves (mill).

Most common machine tools are Mill, Lathe, Saw, Broach, Punch press, Grind, Bore and Stamp break.

CNC

Computer Numerical Control is the automation of machine tools that are operated by precisely programmed commands encoded on a storage medium, as opposed to controlled manually via hand wheels or levers, or mechanically automated via cams alone. Most CNC today is computer numerical control in which computers play an integral part of the control.

3D Printing

All methods work by working in layers, adding material, etc. different to other techniques, which are subtractive. Support is needed because almost all methods could support multi material printing, but it is currently only available with certain top-tier machines.

A method of turning digital shapes into physical objects. Due to its nature, it allows us to accurately control the shape of the product. The drawback is size restraints and materials are often not durable.

While FDM does not seem like the best method for instrument manufacturing, it is one of the cheapest and most universally available methods.

EDM
Electric Discharge Machining.

FDM
Fused Deposition Modeling.

SLA
Stereo Lithography.

SLS
Selective Laser Sintering.

SLM
Selective Laser Melting.

J-P
Jetted Photopolymer (or Polyjet).

EDM Electric Discharge Machining

The basic EDM process is really quite simple. An electrical spark is created between an electrode and a work piece. The spark is visible evidence of the flow of electricity. This electric spark produces intense heat with

temperatures reaching 8000 to 12000 degrees Celsius, melting almost anything.

The spark is very carefully controlled and localized so that it only affects the surface of the material.

The EDM process usually does not affect the heat treat below the surface. With wire EDM the spark always takes place in the dielectric of deionized water. The conductivity of the water is carefully controlled making an excellent environment for the EDM process. The water acts as a coolant and flushes away the eroded metal particles.

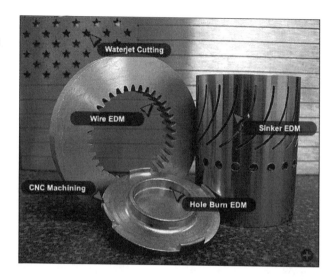

FDM Fused Deposition Modeling

3D printers that run on FDM Technology build parts layer-by-layer by heating thermoplastic material to a semi-liquid state and extruding it according to computer controlled paths.

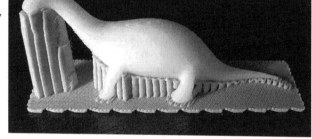

FDM uses two materials to execute a print job: modeling material, which constitutes the finished piece, and support material, which acts as scaffolding. Material filaments are fed from the 3D printer's material bays to the print head, which moves in X and Y coordinates, depositing material to complete each layer before the base moves down the Z axis and the next layer begins.

Once the 3D printer is done building, the user breaks the support material away or dissolves it in detergent and water, and the part is ready to use.

SLA StereoLithograph Apparatus

Stereolithography is an additive fabrication process utilizing a vat of liquid UV-curable photopolymer "resin" and a UV laser to build parts a layer at a time. On each

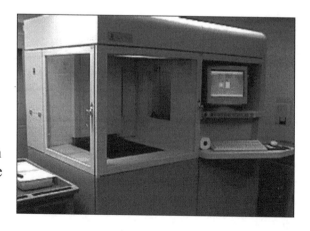

layer, the laser beam traces a part cross-section pattern on the surface of the liquid resin. Exposure to the UV laser light cures, or solidifies the pattern traced on the resin and adheres it to the layer below.

After a pattern has been traced, the SLA's elevator platform descends by a single layer thickness, typically .0019in to .0059in. Then, a resin-filled blade sweeps across the part cross section, re-coating it with fresh material. On this new liquid surface the subsequent layer pattern is traced, adhering to the previous layer.

A complete 3-D part is formed by this process. After building, parts are cleaned of excess resin by immersion in a chemical bath and then cured in a UV oven.

SLS Selective Laser Sintering

Selective laser sintering (SLS) is an additive manufacturing (AM) technique that uses a laser as the power source to sinter (compacting) metal), aiming the laser automatically at points in space defined by a 3D model, binding the material together to create a solid structure.

It is similar to direct metal laser sintering (DMLS); the two are instantiations of the same concept but differ in technical details. Selective laser melting (SLM) uses a comparable concept, but in SLM the material is fully melted rather than sintered,[1] allowing different properties (crystal structure, porosity, and so on). SLS (as well as the other mentioned AM techniques) is a relatively new technology that so far has mainly been used for rapid prototyping and for low-volume production of component parts. Production roles are expanding as the commercialization of AM technology improves.

SLM Selective Laser Melting

Selective laser melting is an additive manufacturing process that uses 3D CAD data as a digital information source and energy in the form of a high-power laser beam, to create three-dimensional metal parts by fusing fine metal powders together. Manufacturing applications in aerospace or medical orthopedics are being pioneered.

The process starts by slicing the 3D CAD file data into layers, usually from 20 to 100 micrometres thick (0.00078740157 to 0.00393700787 in) creating a 2D image of each layer; this file format is the industry standard .stl file used on most layer-based 3D printing or stereolithography technologies.

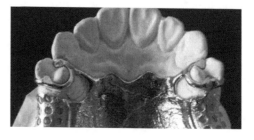

This file is then loaded into a file preparation software package that assigns parameters, values and physical supports that allow the file to be interpreted and built by different types of additive manufacturing machines.

J-P Jetted Photopolymer (or Polyjet)

Photopolymer jetting (or PolyJet) builds prototypes by jetting liquid photopolymer resin from ink-jet style heads. The resin is sprayed from the moving heads, and only the amount of material needed is used.

UV light is simultaneously emitted from the head, which cures each layer of resin immediately after it is applied. The process produces excellent surface finish and feature detail. Photopolymer jetting is used primarily to check form and fit and can handle limited functional tests due to the limited strength of photopolymer resins.

This process offers the unique ability to create prototypes with more than one type of material. For instance, a toothbrush prototype could be composed with a rigid shaft with a rubber-like over-molding for grip.

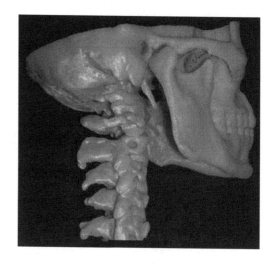

The process works with a variety of proprietary photopolymer resins as opposed to production materials. A tradeoff with this technology is that exposure to ambient heat, humidity, or sunlight can cause dimensional change that can affect tolerance. The process is faster and cleaner than the traditional vat and laser photopolymer processes.

Carbon 3D

The Carbon 3D not only prints composite materials like carbon fiber, but also fiberglass, nylon and PLA. Of course, only one at a time.

The printer employs some pretty nifty advancements, too, including a self-leveling printing bed that clicks into position before each print.

The Carbon 3D is groundbreaking 3D printing technology which is 25 to 100 times faster than currently available commercial PolyJet or SLA machines.

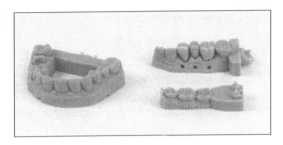

It is a true quantum leap forward for 3D printing speed!

Newton's Law:

1. Every object remains stopped or goes on moving at a steady rate in a straight line unless acted upon by another force. This is the inertia principle.
2. The amount of force needed to make an object change its speed depends on the mass of the object and the amount of acceleration or deceleration required.
3. To every action there is an equal and opposite reaction. When a body is pushed on way by a force, another force pushes back with equal strength.

Polymers:

A polymer is made of one or more large molecules formed from thousands of smaller molecules. Rubber and Wood are natural polymers. Plastics are synthetic (artificially made) polymers.

Speed and Velocity:

Speed is the rate at which a moving object changes position (how far it moves in a fixed time).

Velocity is speed in a particular direction.

If either speed or direction is changed, velocity also changes.

Absorbed

A feature, sketch, or annotation that is contained in another item (usually a feature) in the FeatureManager design tree. Examples are the profile sketch and profile path in a base-sweep, or a cosmetic thread annotation in a hole.

Align

Tools that assist in lining up annotations and dimensions (left, right, top, bottom, and so on). For aligning parts in an assembly.

Alternate position view

A drawing view in which one or more views are superimposed in phantom lines on the original view. Alternate position views are often used to show the range of motion of an assembly.

Anchor point

The end of a leader that attaches to the note, block, or other annotation. Sheet formats contain anchor points for a bill of materials, a hole table, a revision table, and a weldment cut list.

Annotation

A text note or a symbol that adds specific design intent to a part, assembly, or drawing. Specific types of annotations include note, hole callout, surface finish symbol, datum feature symbol, datum target, geometric tolerance symbol, weld symbol, balloon, and stacked balloon. Annotations that apply only to drawings include center mark, annotation centerline, area hatch, and block.

Appearance callouts

Callouts that display the colors and textures of the face, feature, body, and part under the entity selected and are a shortcut to editing colors and textures.

Area hatch

A crosshatch pattern or fill applied to a selected face or to a closed sketch in a drawing.

Assembly

A document in which parts, features, and other assemblies (sub-assemblies) are mated together. The parts and sub-assemblies exist in documents separate from the assembly. For example, in an assembly, a piston can be mated to other parts, such as a connecting rod or cylinder. This new assembly can then be used as a sub-assembly in an assembly of an engine. The extension for a SOLIDWORKS assembly file name is .SLDASM.

Attachment point

The end of a leader that attaches to the model (to an edge, vertex, or face, for example) or to a drawing sheet.

Axis

A straight line that can be used to create model geometry, features, or patterns. An axis can be made in a number of different ways, including using the intersection of two planes.

Balloon

Labels parts in an assembly, typically including item numbers and quantity. In drawings, the item numbers are related to rows in a bill of materials.

Base

The first solid feature of a part.

Baseline dimensions

Sets of dimensions measured from the same edge or vertex in a drawing.

Bend

A feature in a sheet metal part. A bend generated from a filleted corner, cylindrical face, or conical face is a round bend; a bend generated from sketched straight lines is a sharp bends.

Bill of materials

A table inserted into a drawing to keep a record of the parts used in an assembly.

Block

A user-defined annotation that you can use in parts, assemblies, and drawings. A block can contain text, sketch entities (except points), and area hatch, and it can be saved in a file for later use as, for example, a custom callout or a company logo.

Bottom-up assembly

An assembly modeling technique where you create parts and then insert them into an assembly.

Broken-out section

A drawing view that exposes inner details of a drawing view by removing material from a closed profile, usually a spline.

Cavity

The mold half that holds the cavity feature of the design part.

Center mark

A cross that marks the center of a circle or arc.

Centerline

Centerline marks, in phantom font, an axis of symmetry in a sketch or drawing.

Chamfer
Bevels a selected edge or vertex. You can apply chamfers to both sketches and features.

Child
A dependent feature related to a previously built feature. For example, a chamfer on the edge of a hole is a child of the parent hole.

Click-release
As you sketch, if you click and then release the pointer, you are in click-release mode. Move the pointer and click again to define the next point in the sketch sequence.

Click-drag
As you sketch, if you click and drag the pointer, you are in click-drag mode. When you release the pointer, the sketch entity is complete.

Closed profile
Also called a closed contour, it is a sketch or sketch entity with no exposed endpoints, for example, a circle or polygon.

Collapse
The opposite of explode. The collapse action returns an exploded assembly's parts to their normal positions.

Collision Detection
An assembly function that detects collisions between components when components move or rotate. A collision occurs when an entity on one component coincides with any entity on another component.

Component
Any part or sub-assembly within an assembly

Configuration
A variation of a part or assembly within a single document. Variations can include different dimensions, features, and properties. For example, a single part such as a bolt can contain different configurations that vary the diameter and length.

ConfigurationManager
Located on the left side of the SOLIDWORKS window, it is a means to create, select, and view the configurations of parts and assemblies.

Constraint
The relations between sketch entities, or between sketch entities and planes, axes, edges,

or vertices.

Construction geometry

The characteristic of a sketch entity is that the entity is used in creating other geometry but is not itself used in creating features.

Coordinate system

A system of planes used to assign Cartesian coordinates to features, parts, and assemblies. Part and assembly documents contain default coordinate systems; other coordinate systems can be defined with reference geometry. Coordinate systems can be used with measurement tools and for exporting documents to other file formats.

Cosmetic thread

An annotation that represents threads.

Crosshatch

A pattern (or fill) applied to drawing views such as section views and broken-out sections.

Curvature

Curvature is equal to the inverse of the radius of the curve. The curvature can be displayed in different colors according to the local radius (usually of a surface).

Cut

A feature that removes material from a part by such actions as extrude, revolve, loft, sweep, thicken, cavity, and so on.

Dangling

A dimension, relation, or drawing section view that is unresolved. For example, if a piece of geometry is dimensioned, and that geometry is later deleted, the dimension becomes dangling.

Degrees of freedom

Geometry that is not defined by dimensions or relations is free to move. In 2D sketches, there are three degrees of freedom: movement along the X and Y axes, and rotation about the Z axis (the axis normal to the sketch plane). In 3D sketches and in assemblies, there are six degrees of freedom: movement along the X, Y, and Z axes, and rotation about the X, Y, and Z axes.

Derived part

A derived part is a new base, mirror, or component part created directly from an existing part and linked to the original part such that changes to the original part are reflected in the derived part.

Derived sketch
A copy of a sketch, in either the same part or the same assembly that is connected to the original sketch. Changes in the original sketch are reflected in the derived sketch.

Design Library
Located in the Task Pane, the Design Library provides a central location for reusable elements such as parts, assemblies, and so on.

Design table
An Excel spreadsheet that is used to create multiple configurations in a part or assembly document.

Detached drawing
A drawing format that allows opening and working in a drawing without loading the corresponding models into memory. The models are loaded on an as-needed basis.

Detail view
A portion of a larger view, usually at a larger scale than the original view.

Dimension line
A linear dimension line references the dimension text to extension lines indicating the entity being measured. An angular dimension line references the dimension text directly to the measured object.

DimXpertManager
Located on the left side of the SOLIDWORKS window, it is a means to manage dimensions and tolerances created using DimXpert for parts according to the requirements of the ASME Y.14.41-2003 standard.

DisplayManager
The DisplayManager lists the appearances, decals, lights, scene, and cameras applied to the current model. From the DisplayManager, you can view applied content, and add, edit, or delete items.

Document
A file containing a part, assembly, or drawing.

Draft
The degree of taper or angle of a face usually applied to molds or castings.

Drawing
A 2D representation of a 3D part or assembly. The extension for a SOLIDWORKS drawing file name is .SLDDRW.

Drawing sheet

A page in a drawing document.

Driven dimension

Measurements of the model, but they do not drive the model and their values cannot be changed.

Driving dimension

Also referred to as a model dimension, it sets the value for a sketch entity. It can also control distance, thickness, and feature parameters.

Edge

A single outside boundary of a feature.

Edge flange

A sheet metal feature that combines a bend and a tab in a single operation.

Equation

Creates a mathematical relation between sketch dimensions, using dimension names as variables, or between feature parameters, such as the depth of an extruded feature or the instance count in a pattern.

Exploded view

Shows an assembly with its components separated from one another, usually to show how to assemble the mechanism.

Export

Save a SOLIDWORKS document in another format for use in other CAD/CAM, rapid prototyping, web, or graphics software applications.

Extension line

The line extending from the model indicating the point from which a dimension is measured.

Extrude

A feature that linearly projects a sketch to either add material to a part (in a base or boss) or remove material from a part (in a cut or hole).

Face

A selectable area (planar or otherwise) of a model or surface with boundaries that help define the shape of the model or surface. For example, a rectangular solid has six faces.

Fasteners

A SOLIDWORKS Toolbox library that adds fasteners automatically to holes in an assembly.

Feature

An individual shape that, combined with other features, makes up a part or assembly. Some features, such as bosses and cuts, originate as sketches. Other features, such as shells and fillets, modify a feature's geometry. However, not all features have associated geometry. Features are always listed in the FeatureManager design tree.

FeatureManager design tree

Located on the left side of the SOLIDWORKS window, it provides an outline view of the active part, assembly, or drawing.

Fill

A solid area hatch or crosshatch. Fill also applies to patches on surfaces.

Fillet

An internal rounding of a corner or edge in a sketch, or an edge on a surface or solid.

Forming tool

Dies that bend, stretch, or otherwise form sheet metal to create such form features as louvers, lances, flanges, and ribs.

Fully defined

A sketch where all lines and curves in the sketch, and their positions, are described by dimensions or relations, or both, and cannot be moved. Fully defined sketch entities are shown in black.

Geometric tolerance

A set of standard symbols that specify the geometric characteristics and dimensional requirements of a feature.

Graphics area

The area in the SOLIDWORKS window where the part, assembly, or drawing appears.

Guide curve

A 2D or 3D curve is used to guide a sweep or loft.

Handle

An arrow, square, or circle that you can drag to adjust the size or position of an entity (a feature, dimension, or sketch entity, for example).

Helix

A curve defined by pitch, revolutions, and height. A helix can be used, for example, as a path for a swept feature cutting threads in a bolt.

Hem

A sheet metal feature that folds back at the edge of a part. A hem can be open, closed, double, or teardrop.

HLR

(Hidden lines removed) a view mode in which all edges of the model that are not visible from the current view angle are removed from the display.

HLV

(Hidden lines visible) A view mode in which all edges of the model that are not visible from the current view angle are shown gray or dashed.

Import

Open files from other CAD software applications into a SOLIDWORKS document.

In-context feature

A feature with an external reference to the geometry of another component; the in-context feature changes automatically if the geometry of the referenced model or feature changes.

Inference

The system automatically creates (infers) relations between dragged entities (sketched entities, annotations, and components) and other entities and geometry. This is useful when positioning entities relative to one another.

Instance

An item in a pattern or a component in an assembly that occurs more than once. Blocks are inserted into drawings as instances of block definitions.

Interference detection

A tool that displays any interference between selected components in an assembly.

Jog

A sheet metal feature that adds material to a part by creating two bends from a sketched line.

Knit

A tool that combines two or more faces or surfaces into one. The edges of the surfaces must be adjacent and not overlapping, but they cannot ever be planar. There is no difference in the appearance of the face or the surface after knitting.

Layout sketch

A sketch that contains important sketch entities, dimensions, and relations. You reference the entities in the layout sketch when creating new sketches, building new geometry, or positioning components in an assembly. This allows for easier updating of your model because changes you make to the layout sketch propagate to the entire model.

Leader

A solid line from an annotation (note, dimension, and so on) to the referenced feature.

Library feature

A frequently used feature, or combination of features, that is created once and then saved for future use.

Lightweight

A part in an assembly or a drawing has only a subset of its model data loaded into memory. The remaining model data is loaded on an as-needed basis. This improves the performance of large and complex assemblies.

Line

A straight sketch entity with two endpoints. A line can be created by projecting an external entity such as an edge, plane, axis, or sketch curve into the sketch.

Loft

A base, boss, cut, or surface feature created by transitions between profiles.

Lofted bend

A sheet metal feature that produces a roll form or a transitional shape from two open profile sketches. Lofted bends often create funnels and chutes.

Mass properties

A tool that evaluates the characteristics of a part or an assembly such as volume, surface area, centroid, and so on.

Mate

A geometric relationship, such as coincident, perpendicular, tangent, and so on, between parts in an assembly.

Mate reference

Specifies one or more entities of a component to use for automatic mating. When you drag a component with a mate reference into an assembly, the software tries to find other combinations of the same mate reference name and mate type.

Mates folder
A collection of mates that are solved together. The order in which the mates appear within the Mates folder does not matter.

Mirror
(a) A mirror feature is a copy of a selected feature, mirrored about a plane or planar face.
(b) A mirror sketch entity is a copy of a selected sketch entity that is mirrored about a centerline.

Miter flange
A sheet metal feature that joins multiple edge flanges together and miters the corner.

Model
3D solid geometry in a part or assembly document. If a part or assembly document contains multiple configurations, each configuration is a separate model.

Model dimension
A dimension specified in a sketch or a feature in a part or assembly document that defines some entity in a 3D model.

Model item
A characteristic or dimension of feature geometry that can be used in detailing drawings.

Model view
A drawing view of a part or assembly.

Mold
A set of manufacturing tooling used to shape molten plastic or other material into a designed part. You design the mold using a sequence of integrated tools that result in cavity and core blocks that are derived parts of the part to be molded.

Motion Study
Motion Studies are graphical simulations of motion and visual properties with assembly models. Analogous to a configuration, they do not actually change the original assembly model or its properties. They display the model as it changes based on simulation elements you add.

Multibody part
A part with separate solid bodies within the same part document. Unlike the components in an assembly, multibody parts are not dynamic.

Native format

DXF and DWG files remain in their original format (are not converted into SOLIDWORKS format) when viewed in SOLIDWORKS drawing sheets (view only).

Open profile

Also called an open contour, it is a sketch or sketch entity with endpoints exposed. For example, a U-shaped profile is open.

Ordinate dimensions

A chain of dimensions measured from a zero ordinate in a drawing or sketch.

Origin

The model origin appears as three gray arrows and represents the (0,0,0) coordinate of the model. When a sketch is active, a sketch origin appears in red and represents the (0,0,0) coordinate of the sketch. Dimensions and relations can be added to the model origin, but not to a sketch origin.

Out-of-context feature

A feature with an external reference to the geometry of another component that is not open.

Over defined

A sketch is over defined when dimensions or relations are either in conflict or redundant.

Parameter

A value used to define a sketch or feature (often a dimension).

Parent

An existing feature upon which other features depend. For example, in a block with a hole, the block is the parent to the child hole feature.

Part

A single 3D object made up of features. A part can become a component in an assembly, and it can be represented in 2D in a drawing. Examples of parts are bolt, pin, plate, and so on. The extension for a SOLIDWORKS part file name is .SLDPRT.

Path

A sketch, edge, or curve used in creating a sweep or loft.

Pattern

A pattern repeats selected sketch entities, features, or components in an array, which can be linear, circular, or sketch driven. If the seed entity is changed, the other instances in the pattern update.

Physical Dynamics
An assembly tool that displays the motion of assembly components in a realistic way. When you drag a component, the component applies a force to other components it touches. Components move only within their degrees of freedom.

Pierce relation
Makes a sketch point coincident to the location at which an axis, edge, line, or spline pierces the sketch plane.

Planar
Entities that can lie on one plane. For example, a circle is planar, but a helix is not.

Plane
Flat construction geometry. Planes can be used for a 2D sketch, section view of a model, a neutral plane in a draft feature, and others.

Point
A singular location in a sketch, or a projection into a sketch at a single location of an external entity (origin, vertex, axis, or point in an external sketch).

Predefined view
A drawing view in which the view position, orientation, and so on can be specified before a model is inserted. You can save drawing documents with predefined views as templates.

Profile
A sketch entity used to create a feature (such as a loft) or a drawing view (such as a detail view). A profile can be open (such as a U shape or open spline) or closed (such as a circle or closed spline).

Projected dimension
If you dimension entities in an isometric view, projected dimensions are the flat dimensions in 2D.

Projected view
A drawing view projected orthogonally from an existing view.

PropertyManager
Located on the left side of the SOLIDWORKS window, it is used for dynamic editing of sketch entities and most features.

RealView graphics

A hardware (graphics card) support of advanced shading in real time; the rendering applies to the model and is retained as you move or rotate a part.

Rebuild

Tool that updates (or regenerates) the document with any changes made since the last time the model was rebuilt. Rebuild is typically used after changing a model dimension.

Reference dimension

A dimension in a drawing that shows the measurement of an item but cannot drive the model and its value cannot be modified. When model dimensions change, reference dimensions update.

Reference geometry

Includes planes, axes, coordinate systems, and 3D curves. Reference geometry is used to assist in creating features such lofts, sweeps, drafts, chamfers, and patterns.

Relation

A geometric constraint between sketch entities or between a sketch entity and a plane, axis, edge, or vertex. Relations can be added automatically or manually.

Relative view

A relative (or relative to model) drawing view is created relative to planar surfaces in a part or assembly.

Reload

Refreshes shared documents. For example, if you open a part file for read-only access while another user makes changes to the same part, you can reload the new version, including the changes.

Reorder

Reordering (changing the order of) items is possible in the FeatureManager design tree. In parts, you can change the order in which features are solved. In assemblies, you can control the order in which components appear in a bill of materials.

Replace

Substitutes one or more open instances of a component in an assembly with a different component.

Resolved

A state of an assembly component (in an assembly or drawing document) in which it is fully loaded in memory. All the component's model data is available, so its entities can be selected, referenced, edited, and used in mates, and so on.

Revolve

A feature that creates a base or boss, a revolved cut, or revolved surface by revolving one or more sketched profiles around a centerline.

Rip

A sheet metal feature that removes material at an edge to allow a bend.

Rollback

Suppresses all items below the rollback bar.

Section

Another term for profile in sweeps.

Section line

A line or centerline sketched in a drawing view to create a section view.

Section scope

Specifies the components to be left uncut when you create an assembly drawing section view.

Section view

A section view (or section cut) is (1) a part or assembly view cut by a plane, or (2) a drawing view created by cutting another drawing view with a section line.

Seed

A sketch or an entity (a feature, face, or body) that is the basis for a pattern. If you edit the seed, the other entities in the pattern are updated.

Shaded

Displays a model as a colored solid.

Shared values

Also called linked values, these are named variables that you assign to set the value of two or more dimensions to be equal.

Sheet format

Includes page size and orientation, standard text, borders, title blocks, and so on. Sheet formats can be customized and saved for future use. Each sheet of a drawing document can have a different format.

Shell

A feature that hollows out a part, leaving open the selected faces and thin walls on the remaining faces. A hollow part is created when no faces are selected to be open.

Sketch
A collection of lines and other 2D objects on a plane or face that forms the basis for a feature such as a base or a boss. A 3D sketch is non-planar and can be used to guide a sweep or loft, for example.

Smart Fasteners
Automatically adds fasteners (bolts and screws) to an assembly using the SOLIDWORKS Toolbox library of fasteners.

SmartMates
An assembly mating relation that is created automatically.

Solid sweep
A cut sweep created by moving a tool body along a path to cut out 3D material from a model.

Spiral
A flat or 2D helix, defined by a circle, pitch, and number of revolutions.

Spline
A sketched 2D or 3D curve defined by a set of control points.

Split line
Projects a sketched curve onto a selected model face, dividing the face into multiple faces so that each can be selected individually. A split line can be used to create draft features, to create face blend fillets, and to radiate surfaces to cut molds.

Stacked balloon
A set of balloons with only one leader. The balloons can be stacked vertically (up or down) or horizontally (left or right).

Standard 3 views
The three orthographic views (front, right, and top) that are often the basis of a drawing.

StereoLithography
The process of creating rapid prototype parts using a faceted mesh representation in STL files.

Sub-assembly
An assembly document that is part of a larger assembly. For example, the steering mechanism of a car is a sub-assembly of the car.

Suppress

Removes an entity from the display and from any calculations in which it is involved. You can suppress features, assembly components, and so on. Suppressing an entity does not delete the entity; you can un-suppress the entity to restore it.

Surface

A zero-thickness planar or 3D entity with edge boundaries. Surfaces are often used to create solid features. Reference surfaces can be used to modify solid features.

Sweep

Creates a base, boss, cut, or surface feature by moving a profile (section) along a path. For cut sweeps, you can create solid sweeps by moving a tool body along a path.

Tangent arc

An arc that is tangent to another entity, such as a line.

Tangent edge

The transition edge between rounded or filleted faces in hidden lines visible or hidden lines removed modes in drawings.

Task Pane

Located on the right-side of the SOLIDWORKS window, the Task Pane contains SOLIDWORKS Resources, the Design Library, and the File Explorer.

Template

A document (part, assembly, or drawing) that forms the basis of a new document. It can include user-defined parameters, annotations, predefined views, geometry, and so on.

Temporary axis

An axis created implicitly for every conical or cylindrical face in a model.

Thin feature

An extruded or revolved feature with constant wall thickness. Sheet metal parts are typically created from thin features.

TolAnalyst

A tolerance analysis application that determines the effects that dimensions and tolerances have on parts and assemblies.

Top-down design

An assembly modeling technique where you create parts in the context of an assembly by referencing the geometry of other components. Changes to the referenced components propagate to the parts that you create in context.

Triad

Three axes with arrows defining the X, Y, and Z directions. A reference triad appears in part and assembly documents to assist in orienting the viewing of models. Triads also assist when moving or rotating components in assemblies.

Under defined

A sketch is under defined when there are not enough dimensions and relations to prevent entities from moving or changing size.

Vertex

A point at which two or more lines or edges intersect. Vertices can be selected for sketching, dimensioning, and many other operations.

Viewports

Windows that display views of models. You can specify one, two, or four viewports. Viewports with orthogonal views can be linked, which links orientation and rotation.

Virtual sharp

A sketch point at the intersection of two entities after the intersection itself has been removed by a feature such as a fillet or chamfer. Dimensions and relations to the virtual sharp are retained even though the actual intersection no longer exists.

Weldment

A multibody part with structural members.

Weldment cut list

A table that tabulates the bodies in a weldment along with descriptions and lengths.

Wireframe

A view mode in which all edges of the part or assembly are displayed.

Zebra stripes

Simulate the reflection of long strips of light on a very shiny surface. They allow you to see small changes in a surface that may be hard to see with a standard display.

Zoom

To simulate movement toward or away from a part or an assembly.

Index

Model Library
All models designed using SOLIDWORKS

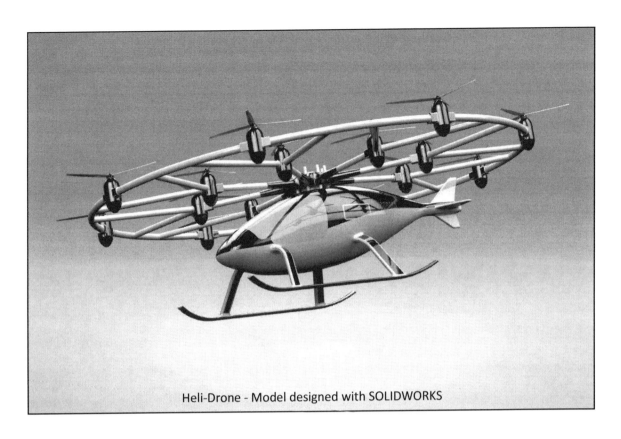

Heli-Drone - Model designed with SOLIDWORKS

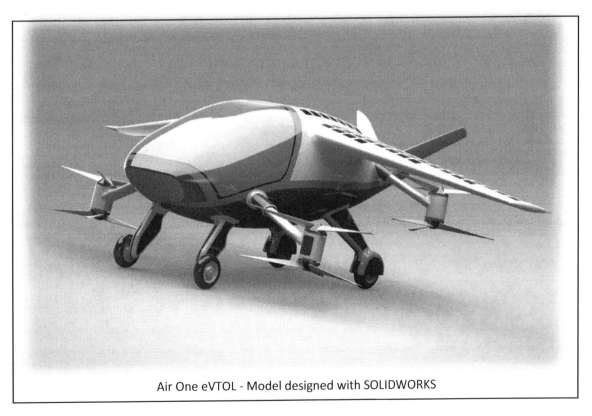

Air One eVTOL - Model designed with SOLIDWORKS

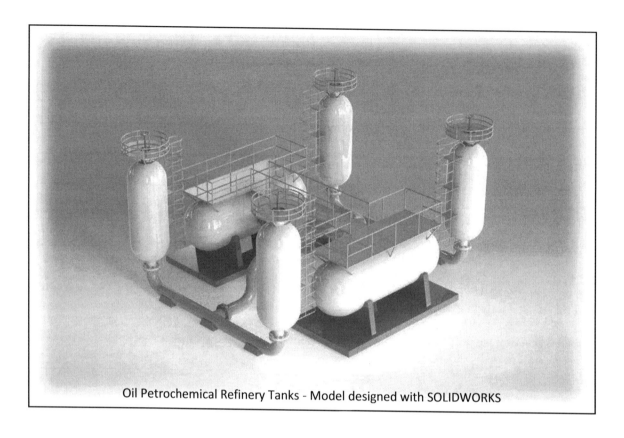

Oil Petrochemical Refinery Tanks - Model designed with SOLIDWORKS

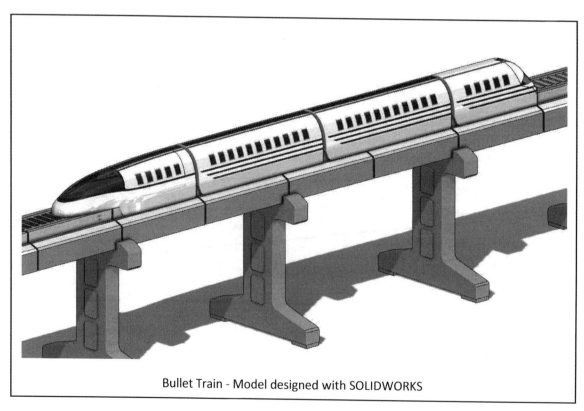

Bullet Train - Model designed with SOLIDWORKS

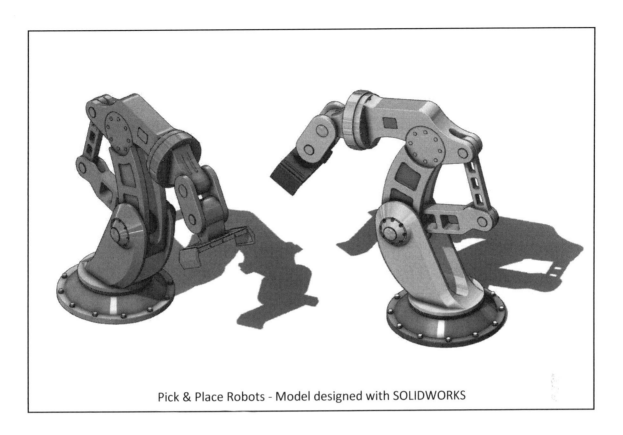

Pick & Place Robots - Model designed with SOLIDWORKS

Spaceship - Model designed with SOLIDWORKS

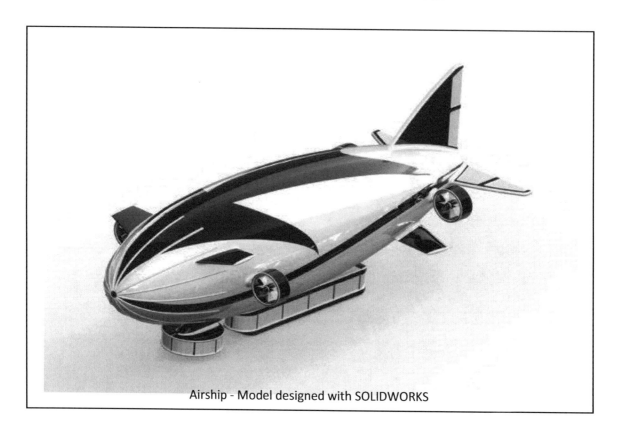

Airship - Model designed with SOLIDWORKS

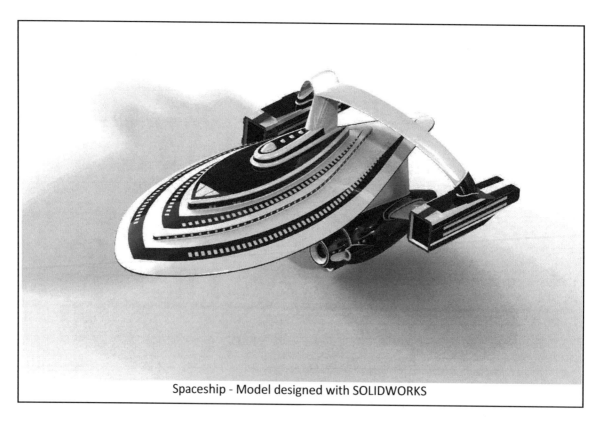

Spaceship - Model designed with SOLIDWORKS

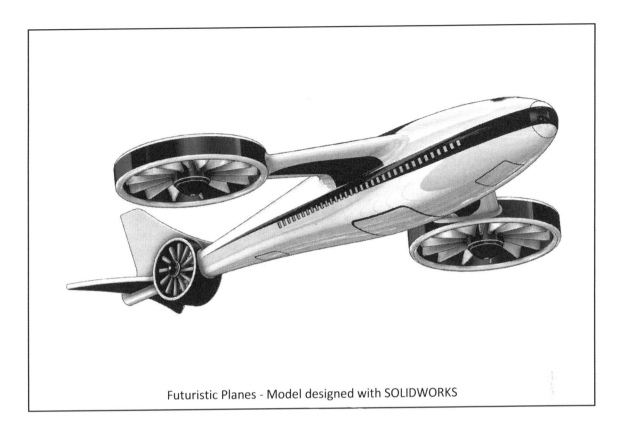

Futuristic Planes - Model designed with SOLIDWORKS

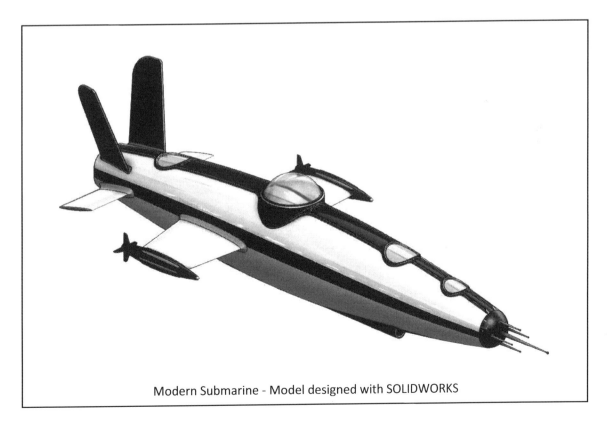

Modern Submarine - Model designed with SOLIDWORKS

Toy Car Designs - Model designed with SOLIDWORKS

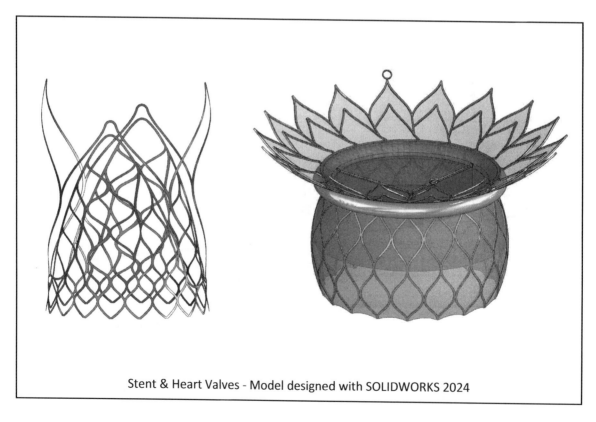

Stent & Heart Valves - Model designed with SOLIDWORKS 2024

Airbus Popup - Model designed with SOLIDWORKS

Airbus Pod - Model designed with SOLIDWORKS

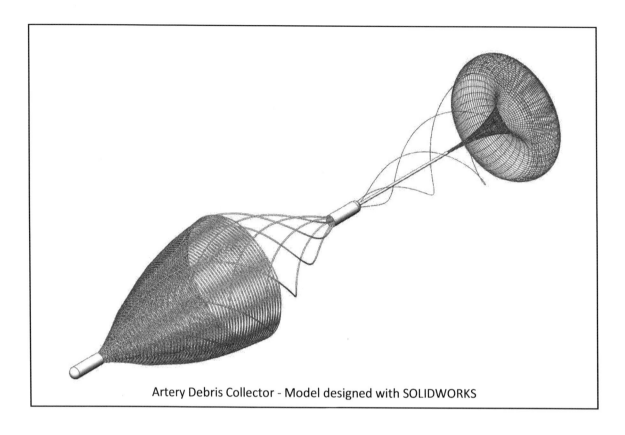

Artery Debris Collector - Model designed with SOLIDWORKS

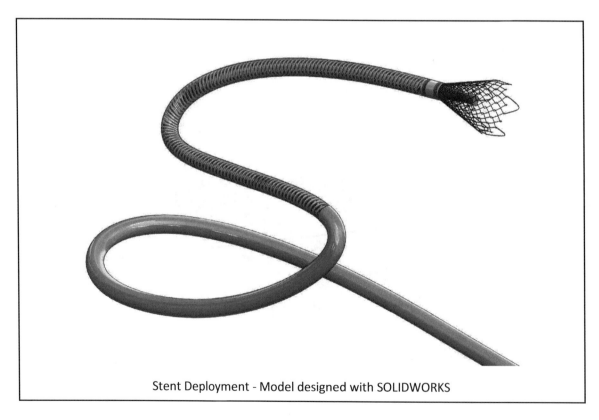

Stent Deployment - Model designed with SOLIDWORKS

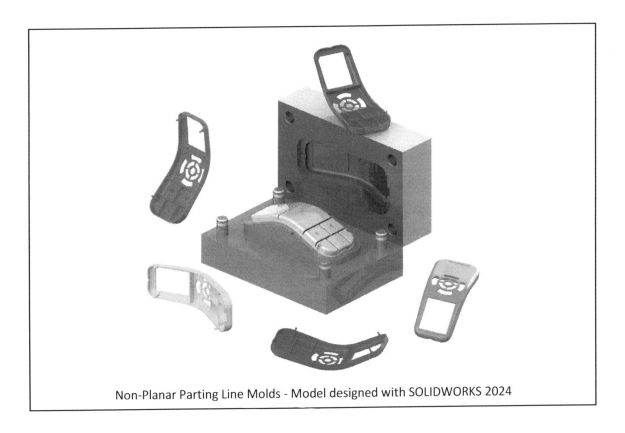

Non-Planar Parting Line Molds - Model designed with SOLIDWORKS 2024

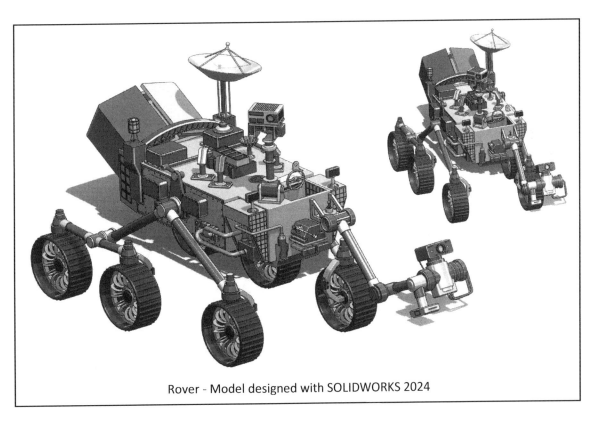

Rover - Model designed with SOLIDWORKS 2024

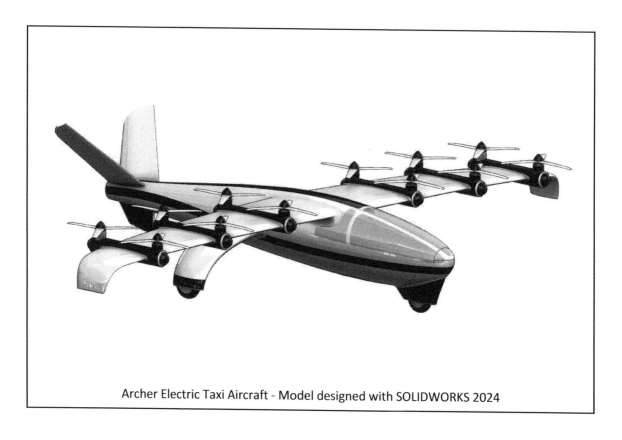

Archer Electric Taxi Aircraft - Model designed with SOLIDWORKS 2024

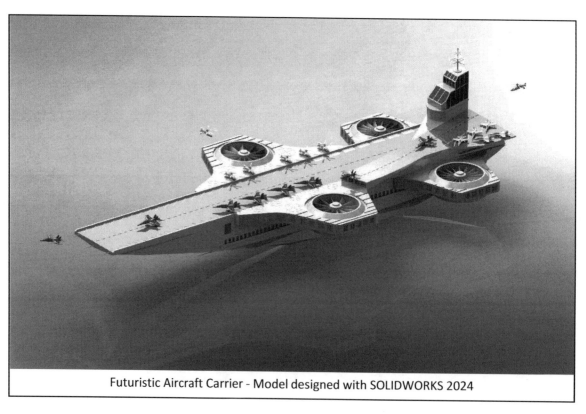

Futuristic Aircraft Carrier - Model designed with SOLIDWORKS 2024

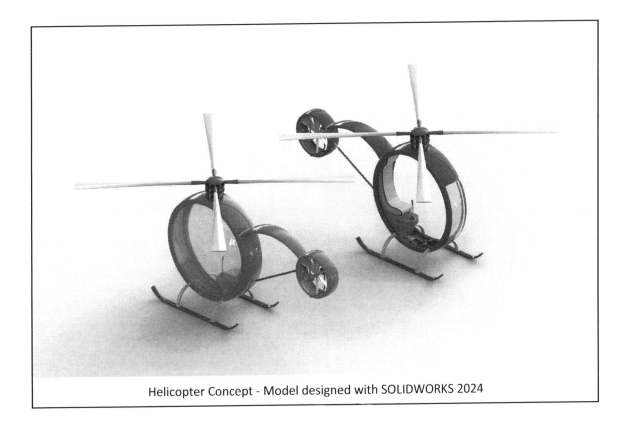

Helicopter Concept - Model designed with SOLIDWORKS 2024

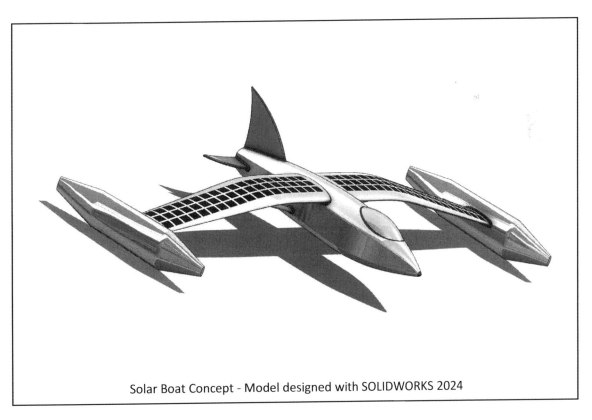

Solar Boat Concept - Model designed with SOLIDWORKS 2024

RC Robo-Snake - Model designed with SOLIDWORKS 2024

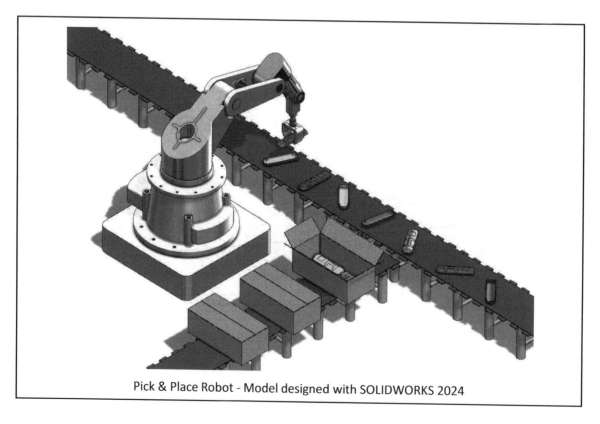

Pick & Place Robot - Model designed with SOLIDWORKS 2024

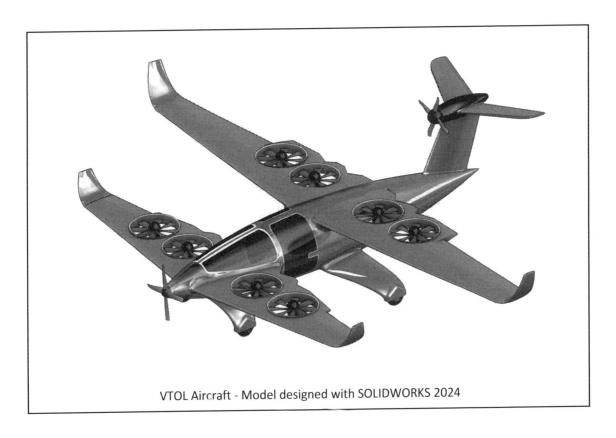

VTOL Aircraft - Model designed with SOLIDWORKS 2024

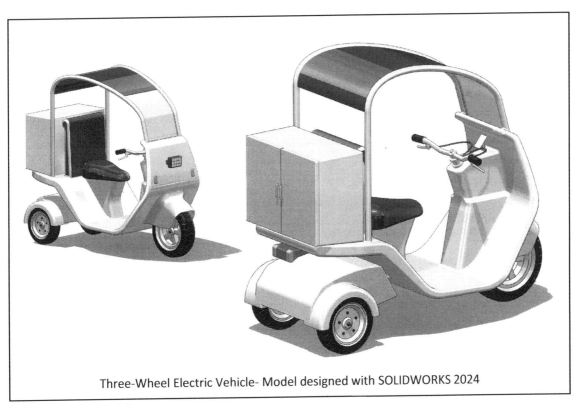

Three-Wheel Electric Vehicle- Model designed with SOLIDWORKS 2024

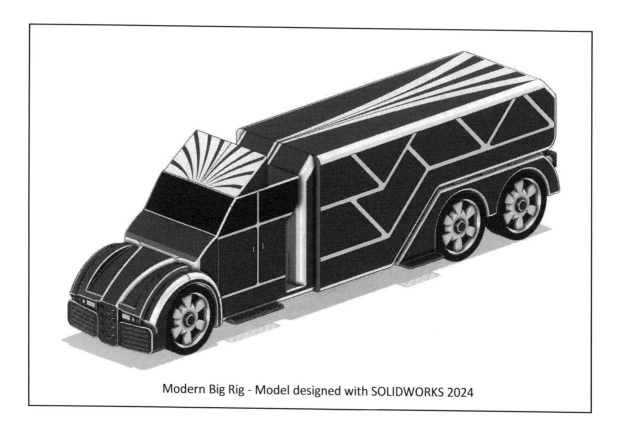

Modern Big Rig - Model designed with SOLIDWORKS 2024

Water Bike - Model designed with SOLIDWORKS

SOLIDWORKS Quick Guide
Command Icons & Toolbars

STANDARD Toolbar

 Creates a new document.

 Opens an existing document.

 Saves an active document.

 Make Drawing from Part/Assembly.

 Make Assembly from Part/Assembly.

 Prints the active document.

 Print preview.

 Cuts the selection & puts it on the clipboard.

 Copies the selection & puts it on the clipboard.

 Inserts the clipboard contents.

 Deletes the selection.

 Reverses the last action.

 Rebuilds the part / assembly / drawing.

 Redo the last action that was undone.

 Saves all documents.

 Edits material.

 Closes an existing document.

 Shows or hides the Selection Filter toolbar.

 Shows or hides the Web toolbar.

 Properties.

 File properties.

 Loads or unloads the 3D instant website add-in.

 Select tool.

 Select the entire document.

 Checks read-only files.

 Options.

 Help.

 Full screen view.

 OK.

 Cancel.

 Magnified selection.

SKETCH TOOLS Toolbar

 Select.

 Sketch.

 3D Sketch.

 Sketches a rectangle from the center.

 Sketches a CenterPoint arc slot.

 Sketches a 3-point arc slot.

 Sketches a straight slot.

 Sketches a CenterPoint straight slot.

 Sketches a 3-point arc.

 Creates sketched ellipses.

Quick Reference Guide to SOLIDWORKS Command Icons & Toolbars

SKETCH TOOLS Toolbar

3D sketch on plane.	Partial ellipses.
Sets up Grid parameters.	Adds a Parabola.
Creates a sketch on a selected plane or face.	Adds a spline.
Equation driven curve.	Sketches a polygon.
Modifies a sketch.	Sketches a corner rectangle.
Copies sketch entities.	Sketches a parallelogram.
Scales sketch entities.	Creates points.
Rotates sketch entities.	Creates sketched centerlines.
Sketches 3-point rectangle from the center.	Adds text to sketch.
Sketches 3-point corner rectangle.	Converts selected model edges or sketch entities to sketch segments.
Sketches a line.	Creates a sketch along the intersection of multiple bodies.
Creates a center point arc: center, start, end.	Converts face curves on the selected face into 3D sketch entities.
Creates an arc tangent to a line.	Mirrors selected segments about a centerline.
Sketches splines on a surface or face.	Fillets the corner of two lines.
Sketches a circle.	Creates a chamfer between two sketch entities.
Sketches a circle by its perimeter.	Creates a sketch curve by offsetting model edges or sketch entities at a specified distance.
Makes a path of sketch entities.	Trims a sketch segment.
Mirrors entities dynamically about a centerline.	Extends a sketch segment.
Insert a plane into the 3D sketch.	Splits a sketch segment.
Instant 2D.	Construction Geometry.
Sketch numeric input.	Creates linear steps and repeat of sketch entities.
Detaches segment on drag.	Creates circular steps and repeat of sketch entitie
Sketch picture.	

SHEET METAL Toolbar

 Add a bend from a selected sketch in a Sheet Metal part.

 Shows flat pattern for this sheet metal part.

 Shows part without inserting any bends.

 Inserts a rib feature to a sheet metal part.

 Create a Sheet Metal part or add material to existing Sheet Metal part.

 Inserts a Sheet Metal Miter Flange feature.

 Folds selected bends.

 Unfolds selected bends.

 Inserts bends using a sketch line.

 Inserts a flange by pulling an edge.

 Inserts a sheet metal corner feature.

 Inserts a Hem feature by selecting edges.

 Breaks a corner by filleting /chamfering it.

 Inserts a Jog feature using a sketch line.

 Inserts a lofted bend feature using 2 sketches.

 Creates inverse dent on a sheet metal part.

 Trims out material from a corner, in a sheet metal part.

 Inserts a fillet weld bead.

 Converts a solid/surface into a sheet metal part.

 Adds a Cross Break feature into a selected face.

 Sweeps an open profile along an open/closed path.

 Adds a gusset/rib across a bend.

 Corner relief.

 Welds the selected corner.

SURFACES Toolbar

 Creates mid surfaces between offset face pairs.

 Patches surface holes and external edges.

 Creates an extruded surface.

 Creates a revolved surface.

 Creates a swept surface.

 Creates a lofted surface.

 Creates an offset surface.

 Radiates a surface originating from a curve, parallel to a plane.

 Knits surfaces together.

 Creates a planar surface from a sketch or a set of edges.

 Creates a surface by importing data from a file.

 Extends a surface.

 Trims a surface.

 Surface flattens.

 Deletes Face(s).

 Replaces Face with Surface.

 Patches surface holes and external edges by extending the surfaces.

 Creates parting surfaces between core & cavity surfaces.

 Inserts ruled surfaces from edges.

WELDMENTS Toolbar

 Creates a weldment feature.

 Creates a structure member feature.

 Adds a gusset feature between 2 planar adjoining faces.

 Creates an end cap feature.

 Adds a fillet weld bead feature.

 Trims or extends structure members.

 Weld bead.

DIMENSIONS/RELATIONS Toolbar

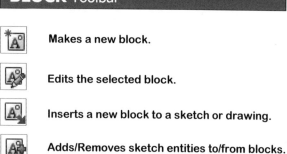

Inserts dimension between two lines.

Creates a horizontal dimension between selected entities.

Creates a vertical dimension between selected entities.

Creates a reference dimension between selected entities.

Creates a set of ordinate dimensions.

Creates a set of Horizontal ordinate

Creates a set of Vertical ordinate dimensions.

Creates a chamfer dimension.

Adds a geometric relation.

Automatically Adds Dimensions to the current sketch.

Displays and deletes geometric relations.

Fully defines a sketch.

Scans a sketch for elements of equal length or radius.

Angular Running dimension.

Display / Delete dimension.

Isolate changed dimension.

Path length dimension.

BLOCK Toolbar

Makes a new block.

Edits the selected block.

Inserts a new block to a sketch or drawing.

Adds/Removes sketch entities to/from blocks.

 Updates parent sketches affected by this block.

 Saves the block to a file.

 Explodes the selected block.

 Inserts a belt.

STANDARD VIEWS Toolbar

 Front view.

 Back view.

 Left view.

 Right view.

 Top view.

 Bottom view.

 Isometric view.

 Trimetric view.

 Dimetric view.

 Normal to view.

 Links all views in the viewport together.

 Displays viewport with front & right

 Displays a 4-view viewport with 1st or 3rd angle of projection.

 Displays viewport with front & top views.

 Displays viewport with a single view.

 View selector.

 New view.

FEATURES Toolbar

 Creates a boss feature by extruding a sketched profile.

 Creates a revolved feature based on profile and angle parameter.

 Creates a cut feature by extruding a sketched profile.

 Creates a cut feature by revolving a sketched profile.

 Thread.

 Creates a cut by sweeping a closed profile along an open or closed path.

 Loft cut.

 Creates a cut by thickening one or more adjacent surfaces.

 Adds a deformed surface by push or pull on points.

 Creates a lofted feature between two or more profiles.

 Creates a solid feature by thickening one or more adjacent surfaces.

 Creates a filled feature.

 Chamfers an edge or a chain of tangent edges.

 Inserts a rib feature.

 Combine.

 Creates a shell feature.

 Applies draft to a selected surface.

 Creates a cylindrical hole.

 Inserts a hole with a pre-defined cross section.

 Puts a dome surface on a face.

 Model break view.

 Applies global deformation to solid or surface bodies.

 Wraps closed sketch contour(s) onto a face.

 Curve Driven pattern.

 Suppresses the selected feature or component.

 Un-suppresses the selected feature or component.

 Flexes solid and surface bodies.

 Intersect.

 Variable Patterns.

 Live Section Plane.

 Mirrors.

 Scale.

 Creates a Sketch Driven pattern.

 Creates a Table-Driven Pattern.

 Inserts a split Feature.

 Hole series.

 Joins bodies from one or more parts into a single part in the context of an assembly.

 Deletes a solid or a surface.

 Instant 3D.

 Inserts apart from file into the active part document.

 Moves/Copies solid and surface bodies or moves graphics bodies.

 Merges short edges on faces.

 Pushes solid / surface model by another solid / surface model.

 Moves face(s) of a solid.

 FeatureWorks Options.

 Linear Pattern.

 Fill Pattern.

 Cuts a solid model with a

 Boundary Boss/Base.

 Boundary Cut.

 Circular Pattern.

 Recognize Features.

 Grid System.

MOLD TOOLS Toolbar

 Extracts core(s) from existing tooling split.

 Constructs a surface patch.

 Moves face(s) of a solid.

 Creates offset surfaces.

 Inserts cavity into a base part.

 Scales a model by a specified factor.

 Applies draft to a selected surface.

 Inserts a split line feature.

 Creates parting lines to separate core & cavity surfaces.

 Finds & creates mold shut-off surfaces.

 Creates a planar surface from a sketch or a set of edges.

 Knits surfaces together.

 Inserts ruled surfaces from edges.

 Creates parting surfaces between core & cavity surfaces.

 Creates multiple bodies from a single body.

 Inserts a tooling split feature.

 Creates parting surfaces between the core & cavity.

 Inserts surface body folders for mold operation.

SELECTION FILTERS Toolbar

 Turns selection filters on and off.

 Clears all filters.

 Selects all filters.

 Inverts current selection.

 Allows selection of edges only.

 Allows selection filter for vertices only.

 Allows selection of faces only.

 Adds filter for Surface Bodies.

 Adds filter for Solid Bodies.

 Adds filter for Axes.

 Adds filter for Planes.

 Adds filter for Sketch Points.

 Allows selection for sketch only.

 Adds filter for Sketch Segments.

 Adds filter for Midpoints.

 Adds filter for Center Marks.

 Adds filter for Centerline.

 Adds filter for Dimensions and Hole Callouts.

 Adds filter for Surface Finish Symbols.

 Adds filter for Geometric Tolerances.

 Adds filter for Notes / Balloons.

 Adds filter for Weld Symbols.

 Adds filter for Weld beads.

 Adds filter for Datum Targets.

 Adds filter for Datum feature only.

 Adds filter for blocks.

 Adds filter for Cosmetic Threads.

Adds filter for Dowel pin symbols.

 Adds filter for connection points.

 Adds filter for routing points.

SOLIDWORKS Add-Ins Toolbar

 Loads/unloads CircuitWorks add-in.

 Loads/unloads the Design Checker add-in.

 Loads/unloads the Visualize add-in.

 Loads/unloads the Scan-to-3D add-in.

 Loads/unloads the SOLIDWORKS Motions add-in.

 Loads/unloads the SOLIDWORKS Routing add-in.

 Loads/unloads the SOLIDWORKS Simulation add-in.

 Loads/unloads the SOLIDWORKS Toolbox add-in.

 Loads/unloads the SOLIDWORKS TolAnalysis add-in.

 Loads/unloads the SOLIDWORKS Flow Simulation add-in.

 Loads/unloads the SOLIDWORKS Plastics add-in.

 Loads/unloads the SOLIDWORKS MBD SNL license.

FASTENING FEATURES Toolbar

 Creates a parameterized mounting boss.

 Creates a parameterized snap hook.

 Creates a groove to mate with a hook feature.

 Uses sketch elements to create a vent for air flow.

 Creates a lip/groove feature.

SCREEN CAPTURE Toolbar

 Copies the current graphics window to the clipboard.

 Records the current graphics window to an AVI file.

 Stops recording the current graphics window to an AVI file.

EXPLODE LINE SKETCH Toolbar

 Adds a route line that connects entities.

 Adds a jog to the route lines.

LINE FORMAT Toolbar

 Changes layer properties.

 Changes the current document layer.

 Changes line color.

 Changes line thickness.

 Changes line style.

 Hides / Shows a hidden edge.

 Changes line display mode.

Did you know??

* Ctrl+Q will force a rebuild on all features of a part.

* Ctrl+B will rebuild the feature being worked on and its dependents.

2D-To-3D Toolbar

 Makes a Front sketch from the selected entities.

 Makes a Top sketch from the selected entities.

 Makes a Right sketch from the selected entities.

Makes a Left sketch from the selected entities.

Makes a Bottom sketch from the selected entities.

Makes a Back sketch from the selected entities.

Makes an Auxiliary sketch from the selected entities.

Creates a new sketch from the selected entities.

Repairs the selected sketch.

Aligns a sketch to the selected point.

Creates an extrusion from the selected sketch segments, starting at the selected sketch point.

Creates a cut from the selected sketch segments, optionally starting at the selected sketch point.

ALIGN Toolbar

Aligns the left side of the selected annotations with the leftmost annotation.

Aligns the right side of the selected annotations with the rightmost annotation.

Aligns the top side of the selected annotations with the topmost annotation.

Aligns the bottom side of the selected annotations with the lowermost annotation.

Evenly spaces the selected annotations horizontally.

Evenly spaces the selected annotations vertically.

Centrally aligns the selected annotations horizontally.

Centrally aligns the selected annotations vertically.

Compacts the selected annotations horizontally.

Compacts the selected annotations vertically.

Creates a group from the selected items.

Deletes the grouping between these items.

Aligns & groups selected dimensions along a line or an arc.

Aligns & groups dimensions at uniform distances.

Evenly spaces selected dimensions.

Aligns collinear selected dimensions.

Aligns stagger selected dimensions.

SOLIDWORKS MBD Toolbar

Captures 3D view.

Manages 3D PDF templates.

Creates shareable 3D PDF presentations.

Toggles dynamic annotation views.

MACRO Toolbar

Runs a Macro.

Stops Macro recorder.

Records (or pauses recording of) actions to create a Macro.

Launches the Macro Editor and begins editing a new macro.

Opens a Macro file for editing.

Creates a custom macro.

SMARTMATES icons

Concentric & Coincident 2 circular edges.

Concentric 2 cylindrical faces.

Coincident 2 linear edges.

Coincident 2 planar faces.

Coincident 2 vertices.

Coincident 2 origins or coordinate systems.

TABLE Toolbar

 Adds a hole table of selected holes from a specified origin datum.

 Adds a Bill of Materials.

 Adds a revision table.

 Displays a Design table in a drawing.

 Adds a weldments cuts list table.

 Adds an Excel based Bill of Materials.

 Adds a weldment cut list table.

REFERENCE GEOMETRY Toolbar

 Adds a reference plane.

 Creates an axis.

 Creates a coordinate system.

 Adds the center of mass.

 Specifies entities to use as references using SmartMates.

SPLINE TOOLS Toolbar

 Inserts a point to a spline.

 Displays all points where the concavity of selected spline changes.

 Displays minimum radius of selected spline.

 Displays curvature combs of selected spline.

 Reduces numbers of points in a selected spline.

 Adds a tangency control.

 Adds a curvature control.

 Adds a spline based on selected sketch entities & edges.

 Displays the spline control polygon.

ANNOTATIONS Toolbar

 Inserts a note.

 Inserts a surface finish symbol.

 Inserts a new geometric tolerancing symbol.

 Attaches a balloon to the selected edge or face.

 Adds balloons for all components in selected view.

 Inserts a stacked balloon.

 Attaches a datum feature symbol to a selected edge / detail.

 Inserts a weld symbol on the selected edge / face / vertex.

 Inserts a datum target symbol and / or point attached to a selected edge / line.

 Selects and inserts block.

 Inserts annotations & reference geometry from the part / assembly into the selected.

 Adds center marks to circles on model.

 Inserts a Centerline.

 Inserts a hole callout.

 Adds a cosmetic thread to the selected cylindrical feature.

 Inserts a Multi-Jog leader.

 Selects a circular edge or an arc for Dowel pin symbol insertion.

 Adds a view location symbol.

 Inserts latest version symbol.

 Adds cross hatch patterns or solid fill.

 Adds a weld bead caterpillar on an edge.

 Adds a weld symbol on a selected entity.

 Inserts a revision cloud.

 Inserts a magnetic line.

 Hides/shows annotation.

DRAWINGS Toolbar

 Updates the selected view to the model's current stage.

 Creates a detail view.

 Creates a section view.

 Inserts an Alternate Position view.

 Unfolds a new view from an existing view.

 Generates a standard 3-view drawing (1st or 3rd angle).

 Inserts an auxiliary view of an inclined surface.

 Adds an Orthogonal or Named view based on an existing part or assembly.

 Adds a Relative view by two orthogonal faces or planes.

 Adds a Predefined orthogonal projected or Named view with a model.

 Adds an empty view.

 Adds vertical break lines to selected view.

 Crops a view.

 Creates a Broken-out section.

QUICK SNAP Toolbar

 Snap to points.

 Snap to center points.

 Snap to midpoints.

 Snap to quadrant points.

 Snap to intersection of 2 curves.

 Snap to nearest curve.

 Snap tangent to curve.

 Snap perpendicular to curve.

 Snap parallel to line.

 Snap horizontally / vertically to points.

 Snap horizontally / vertically.

 Snap to discrete line lengths.

 Snap to angle.

LAYOUT Toolbar

 Creates the assembly layout sketch.

 Sketches a line.

 Sketches a corner rectangle.

 Sketches a circle.

 Sketches a 3-point arc.

 Rounds a corner.

 Trims or extends a sketch.

 Adds sketch entities by offsetting faces, edges and curves.

 Mirrors selected entities about a centerline.

 Adds a relation.

 Creates a dimension.

 Displays / Deletes geometric relations.

 Makes a new block.

 Edits the selected block.

 Inserts a new block to the sketch or drawing.

 Adds / Removes sketch entities to / from a block.

 Saves the block to a file.

 Explodes the selected block.

 Creates a new part from a layout sketch block.

 Positions 2 components relative to one another.

Projects sketch onto selected surface.

 Inserts a split line feature.

 Creates a composite curve from selected edges, curves and sketches.

 Creates a curve through free points.

 Creates a 3D curve through reference points.

 Helical curve defined by a base sketch and shape parameters.

 Displays a view in the selected orientation.

 Reverts to previous view.

 Redraws the current window.

 Zooms out to see entire model.

 Zooms in by dragging a bounding box.

 Zooms in or out by dragging up or down.

 Zooms to fit all selected entities.

 Dynamic view rotation.

 Scrolls view by dragging.

 Displays image in wireframe mode.

 Displays hidden edges in gray.

 Displays image with hidden lines removed.

 Controls the visibility of planes.

 Controls the visibility of axis.

 Controls the visibility of parting lines.

 Controls the visibility of temporary axis.

 Controls the visibility of origins.

 Controls the visibility of coordinate systems.

 Controls the visibility of reference curves.

 Controls the visibility of sketches.

 Controls the visibility of 3D sketch planes.

 Controls the visibility of 3D sketch.

 Controls the visibility of all annotations.

 Controls the visibility of reference points.

 Controls the visibility of routing points.

 Controls the visibility of lights.

 Controls the visibility of cameras.

 Controls the visibility of sketch relations.

 Changes the display state for the current configuration.

 Rolls the model view.

 Turns the orientation of the model view.

 Dynamically manipulate the model view in 3D to make selection.

 Changes the display style for the active view.

 Displays a shade view of the model with its edges.

 Displays a shade view of the model.

 Toggles between draft quality & high quality HLV.

 Cycles through or applies a specific scene.

 Views the models through one of the model's cameras.

 Displays a part or assembly w/different colors according to the local radius of curvature.

 Displays zebra stripes.

 Displays a model with hardware accelerated shades.

 Applies a cartoon effect to model edges & faces.

 Views simulations symbols.

TOOLS Toolbar

 Calculates the distance between selected items.

 Adds or edits equation.

 Calculates the mass properties of the model.

 Checks the model for geometry errors.

 Inserts or edits a Design Table.

 Evaluates section properties for faces and sketches that lie in parallel planes.

 Reports Statistics for this Part/Assembly.

 Deviation Analysis.

 Runs the SimulationXpress analysis wizard Powered by SOLIDWORKS Simulation.

 Checks the spelling.

 Import diagnostics.

 Runs the DFMXpress analysis wizard.

 Runs the SOLIDWORKS FloXpress analysis wizard.

ASSEMBLY Toolbar

 Creates a new part & inserts it into the assembly.

 Adds an existing part or sub-assembly to the assembly.

 Creates a new assembly & inserts it into the assembly.

 Turns on/off large assembly mode for this document.

 Hides / shows model(s) associated with the selected model(s).

 Toggles the transparency of components.

 Changes the selected components to suppressed or resolved.

 Inserts a belt.

 Toggles between editing part and assembly.

 Smart Fasteners.

 Positions two components relative to one another.

 External references will not be created.

 Moves a component.

 Rotates an un-mated component around its center point.

 Replaces selected components.

 Replaces mate entities of mates of the selected components on the selected Mates group.

 Creates a New Exploded view.

 Creates or edits explode line sketch.

 Interference detection.

 Shows or Hides the Simulation toolbar.

 Patterns components in one or two linear directions.

 Patterns components around an axis.

 Sets the transparency of the components other than the one being edited.

 Sketch driven component pattern.

 Pattern driven component pattern.

 Curve driven component pattern.

 Chain driven component pattern.

 SmartMates by dragging & dropping components.

 Checks assembly hole alignments.

 Mirrors subassemblies and parts.

To add or remove an icon to or from the toolbar, first select:

Tools/Customize/Commands

Next, select a **Category**, click a button to see its description and then drag/drop the command icon into any toolbar.

SOLIDWORKS Quick-Guide©
STANDARD Keyboard Shortcuts

Rotate the model

* Horizontally or Vertically: _____ Arrow keys
* Horizontally or Vertically 90°: _____ Shift + Arrow keys
* Clockwise or Counterclockwise: _____ Alt + left or right Arrow
* Pan the model: _____ Ctrl + Arrow keys
* Zoom in: _____ Z (shift + Z or capital Z)
* Zoom out: _____ z (lower case z)
* Zoom to fit: _____ F
* Previous view: _____ Ctrl+Shift+Z

View Orientation

* View Orientation Menu: _____ Space bar
* Front: _____ Ctrl+1
* Back: _____ Ctrl+2
* Left: _____ Ctrl+3
* Right: _____ Ctrl+4
* Top: _____ Ctrl+5
* Bottom: _____ Ctrl+6
* Isometric: _____ Ctrl+7

Selection Filter & Misc.

* Filter Edges: _____ e
* Filter Vertices: _____ v
* Filter Faces: _____ x
* Toggle Selection filter toolbar: _____ F5
* Toggle Selection Filter toolbar (on/off): _____ F6
* New SOLIDWORKS document: _____ F1
* Open Document: _____ Ctrl+O
* Open from Web folder: _____ Ctrl+W
* Save: _____ Ctrl+S
* Print: _____ Ctrl+P
* Magnifying Glass Zoom _____ g
* Switch between the SOLIDWORKS documents _____ Ctrl + Tab

SOLIDWORKS Sample Customized Hot Keys

Function Keys

F1	SW-Help
F2	2D Sketch
F3	3D Sketch
F4	Modify
F5	Selection Filters
F6	Move (2D Sketch)
F7	Rotate (2D Sketch)
F8	Measure
F9	Extrude
F10	Revolve
F11	Sweep
F12	Loft

Sketch

C	Circle
P	Polygon
E	Ellipse
O	Offset Entities
Alt + C	Convert Entities
M	Mirror
Alt + M	Dynamic Mirror
Alt + F	Sketch Fillet
T	Trim
Alt + X	Extend
D	Smart Dimension
Alt + R	Add Relation
Alt + P	Plane
Control + F	Fully Define Sketch
Control + Q	Exit Sketch

Quick-Guide, Part of SOLIDWORKS Basic Tools, Intermediate Skills and Advanced Techniques

SOLIDWORKS Quick-Guide by Paul Tran – Sr. Certified SOLIDWORKS Instructor
Issue 18 / Jan-2024 - Printed in The United State of America – All Rights Reserved